Springer Theses

Recognizing Outstanding Ph.D. Research

For further volumes:
http://www.springer.com/series/8790

Aims and Scope

The series "Springer Theses" brings together a selection of the very best Ph.D. theses from around the world and across the physical sciences. Nominated and endorsed by two recognized specialists, each published volume has been selected for its scientific excellence and the high impact of its contents for the pertinent field of research. For greater accessibility to non-specialists, the published versions include an extended introduction, as well as a foreword by the student's supervisor explaining the special relevance of the work for the field. As a whole, the series will provide a valuable resource both for newcomers to the research fields described, and for other scientists seeking detailed background information on special questions. Finally, it provides an accredited documentation of the valuable contributions made by today's younger generation of scientists.

Theses are accepted into the series by invited nomination only and must fulfill all of the following criteria

- They must be written in good English.
- The topic should fall within the confines of Chemistry, Physics, Earth Sciences, Engineering and related interdisciplinary fields such as Materials, Nanoscience, Chemical Engineering, Complex Systems and Biophysics.
- The work reported in the thesis must represent a significant scientific advance.
- If the thesis includes previously published material, permission to reproduce this must be gained from the respective copyright holder.
- They must have been examined and passed during the 12 months prior to nomination.
- Each thesis should include a foreword by the supervisor outlining the significance of its content.
- The theses should have a clearly defined structure including an introduction accessible to scientists not expert in that particular field.

Giovanna Miritello

Temporal Patterns of Communication in Social Networks

Doctoral Thesis accepted by
the Universidad Carlos III Madrid, Spain

Springer

Author
Dr. Giovanna Miritello
Department of Mathematics
Universidad Carlos III de Madrid
Leganes
Spain

Supervisor
Prof. Esteban Moro
Department of Mathematics
Universidad Carlos III de Madrid
Leganes
Spain

Co-Supervisor
Mr. Rubén Lara
Telefonica Research Ronda de la Comunicacion
Madrid
Spain

ISSN 2190-5053 ISSN 2190-5061 (electronic)
ISBN 978-3-319-00109-8 ISBN 978-3-319-00110-4 (eBook)
DOI 10.1007/978-3-319-00110-4
Springer Cham Heidelberg New York Dordrecht London

Library of Congress Control Number: 2013934544

Printed on acid-free paper

Springer is part of Springer Science+Business Media (www.springer.com)

To my wonderful family

Supervisor's Foreword

While some properties of modern society can be explained by static structures, the ability to communicate faster and further than ever challenges the traditional stationary approach in the study of networks of human communication. Against the idea that ever-lasting and/or permanent ties can describe social interactions, the availability of electronic data of how humans interact has revealed that communication is a far more complex dynamical process shaped by cognitive, attention, or time restrictions; i.e., social networks are in fact temporal networks. While many works in the literature have concentrated in specific problems of temporal networks, there was no understanding of what is the general evolving description of social networks.

Giovanna's thesis advances our current knowledge of what are the temporal and dynamical patterns of human interactions in large social networks and of their impact in processes like information spreading, opinion dynamics, social influence, group formation, and viral marketing, for example. Using a massive dataset of phone call communication traces and a wide range of detection, analysis, and modeling tools, Giovanna was able to characterized the different dynamical process that operate in a social network at very different time scales.

At the smallest time scale, Giovanna found that the bursty patterns of human communication and the fact that humans communicate in groups (conversations) are the most important features to understand the spreading of information in social networks. More broadly, Giovanna solved the entanglement problem between bursty communication and/or tie formation and find the dynamical social strategies that human follow to form and destroy ties along time. This led to the remarkable result of finding that people display very different (from stable to exploratory) social dynamical strategies and their impact and dependence on information accessibility, gender, and/or age of individuals.

The picture of social networks that emerges from Giovanna's work is that of an ever-changing object in which both the structure and dynamics of interactions between humans are tightly coupled and have (both) a deep impact on processes happening in society. In this sense, Giovanna's thesis has changed the way we look at social networks by exhibiting its dynamical nature, the way to measure and/or model its temporal properties, and its implications in social phenomena.

Madrid, January 2013 Prof. Esteban Moro

Preface

This thesis has been a joint project between Universidad Carlos III de Madrid and Telefónica Research (Spain). Specifically, the research has been conducted at the Grupo Interdisciplinar de Sistemas Complejos (GISC), group of Universidad Carlos III, and at the Analytics and Data Mining and User Modelling teams of Telefónica Research.

The main interest of this research has been in understanding and characterizing large networks of human interactions as continuously changing objects, in which members appear and disappear over time and in which interactions are characterized by temporal correlations and inhomogeneities. This constitutes a very challenging and novel topic. In fact, although many real social networks are *temporal* or *dynamical* networks, with elements and properties continuously changing over time, traditional approaches to social network analysis are essentially static: ties (and tie weights) are given by the aggregated activity observed in a given time period, nodes and ties are considered persistent over time, temporal inhomogeneities and correlations between interaction events are neglected, etc. Within this frame, therefore, the time dimension of human behavior has typically been projected out.

Although much effort has been devoted in the last years to characterize the temporal patterns of human interaction, a general understanding of how to model dynamically real social networks is still missing. In this thesis, we contribute to advancing the state of the art in this area by investigating the *instantaneous*, instead than the *aggregated*, contact network and by analyzing the role of temporal activity patterns of human interaction in the description and modeling of real social networks. Specifically, we investigated the role that topological and, in particular, temporal patterns of human interaction play in three main topics of social network analysis and data mining: the characterization of time (or attention) allocation in social networks, the prediction of link decay and/or persistence and the analysis and modeling of information spreading phenomena.

To this end, we have analyzed large anonymized datasets of phone call communication traces (Call Detail Records or CDR) over long periods of time. Access to these observations was granted by Telefónica Research. The availability of empirical data about such massive networks allowed us to analyze and measure global features of human behavior and interaction, and to characterize phenomena and tendencies that might be invisible at small scale. At the same time, the fine-grained resolution of the datasets we had access to and the fact that they cover a large sample of the population, ensure the significance and universality of our results.

The findings that emerge from our research indicate that the observed inhomogeneities and correlations of human temporal patterns of interactions significantly affect the current view of social networks, shifting from a very steady to a highly complex dynamical entity. Temporal patterns of communication are essential not only for a better characterization of the inherent properties of human behavior, but also, and more importantly, for the understanding and modeling of all those phenomena which are triggered by the way in which people communicate and behave. Examples are diffusion of epidemics, information spreading, opinion and influence phenomena, and group formation. Our results indicate the necessity to incorporate temporal patterns of communication in the analysis of social networks: since structure and dynamics are tightly coupled, the analysis and modeling of human behavior has to factor in both.

The work of this thesis combines data mining, analysis of large datasets, theoretical modeling, and simulations and experiments on empirical data. In addition, our work has a wide range of applications in many business sectors. In particular, at Telefónica Research, part of our techniques and findings have been successfully applied to areas such as social networks analysis, modeling human influence, and customer segmentation and targeting in viral marketing campaigns.

We believe this work has made a contribution to understanding and modeling real social networks and we are confident that it will encourage further research in this field.

Financing for the work has been provided in part by Ministerio de Ciencia e Innovación (Spain) through grants i-MATH and MOSAICO and by Telefónica Research in form of the two-years fellowship *Becas de Formación de Doctores Telefónica I+D y Universidad Carlos III de Madrid*, and two additional years of research contract. Telefónica Research also provided access to anonymized datasets on which a critical part of the research presented in this thesis was based.

Acknowledgments

Correction does much, but encouragement does more.

—Johann Wolfgang von Goethe

The contributions of many different people, in their different ways, have made this thesis possible and it is with great pleasure that I would like to express my gratitude to all of them. First and foremost, I would like to thank my supervisor Esteban Moro, whose passion and enthusiasm for research were contagious and motivational for me. I have appreciated all his contributions of time, effort, advices, and ideas to make my Ph.D. experience very stimulating and productive since the first moment. He is without any doubt one of my best role models for teacher, mentor, and scientist.

A very special thanks goes to my second supervisor, Rubén Lara. He has always provided me encouraging and constructive feedbacks. His help and his spirit of positive and proactive cooperation have been essential throughout my Ph.D. I have learned a lot from his experience and expertise in both academic and industry research and he is for me a prime example of what a perfect team leader should be.

I could not have asked for better supervisors. Their high scientific standards, hard work, and ability to select and approach compelling research problems, together with their constructive criticism and complicity, have been tremendous assets to my professional and personal growth. It was a pleasure to work with them and I look forward to many other fruitful collaborations in the future.

I am especially grateful to Nuria Oliver, Scientific Director of the Multimedia and Data Mining and User Modelling Research Teams at Telefónica Research, for her feedbacks and good will and for her constant belief and strong interest in my research project and in my career.

I would like to show my gratitude to the members of the Department of Mathematics at Universidad Carlos III de Madrid. A special thanks goes to the head of the department, Paco Marcellán, and to all the members of Complex Systems Interdisciplinary Group (GISC), for their help and advices and for providing a nice atmosphere within the department and a stimulating environment in which to learn and grow. I would also like to thank the director of the Ph.D. program, José Manuel Rodríguez, and the secretaries of the department, Alberto Calvo and Natalia Delgado, for patiently answering all my e-mails and requests and for helping me with the last-time bureaucracy of the thesis.

All my colleagues from User Modelling group at Telefónica Research supported me in my research and work. I want to thank all of them for their continuous interest in my project, help, and valuable hints. Especially, I am obliged to Pedro, David M., Susana, Ana, Miguel Angel, Alberto, Rocío, and Rafa for all their personal and professional help that they have extended to me throughout these years. Working at Telefónica Research also gave me the opportunity to come across many funny and great young scientists from all over the world. Felipe, Mayank, Luigi, Victor, Jesus, Andreea, Cristina, and Alberto deserve a special mention, for providing a very friendly environment, unforgettable coffee breaks, and valuable suggestions.

I would like to thank all people with whom I collaborated. A special gratitude goes to Sam Roberts and Robin Dunbar. It was a great pleasure for me to work with them. Special thanks are also given to Manuel Cebrián, whose ideas and insights significantly contributed to part of this research.

I also thank my thesis committee members Pablo Rodriguez Rodriguez, Albert Díaz-Guilera, Jari Sarämaki, and Rodolfo Cuerno Rejado for reading the dissertation and making suggestions for changes and improvements.

These years of Ph.D. were made enjoyable and pleasant in large part due to the funny and good friends who became part of my life. Special thanks go to Jelena, Joka, Pivo, Juanma, Svetozar, Kenier, Daniele, Carlos, Matteo, Cristina, Silvia, Eli, Marisa, Emma, Nicola, Marco, Paloma, Zoë, Lara, Edoardo, Luis, Leena, Richi, Fran, and Maria.

Of course no acknowledgments would be complete without giving thanks to my friends from Sicily, especially to Rita G., Gaetano S., Concetta, Gaetano F., Danilo, Salvo V., Salvo G., Paola, Cristina, and Roberta for their support and interest in my research and to Salvo B., Nadia, and Roberto for their continuous encouragement and for the hilarious and sarcastic comments, as successful postdocs, during the very last period of the thesis. Special thanks are given to my best friends Claudia, Enrica, Valentina, and Laura. Despite the geographical distance, their company, love, and advices were always consistently timely and incredibly useful.

I owe my biggest thanks to Dario, for his love, patience, and encouragement. His insightful comments and daily questions about my work were fundamental for me in all the time of research for and writing of this thesis. Not only he is my biggest fan and supporter but also, as a good physicist, he was always the first reviewer of my work.

Lastly, and most important, I would like to thank my family, for endless love, encouragement, advices, and support in all my pursuits. I am especially thankful to my parents, who are the compass of my life, to my sisters Maria and Carla and to my brother Nicola, who are always there to celebrate my accomplishments and to help overcoming my difficulties.

Contents

Chapter 1
Introduction and Motivations

Science advances whenever we can take something that was once invisible and make it visible; and this is now taking place with regard to social networks and social processes.

—Jon Kleinberg

Uncovering the patterns that characterize human behavior is not only one of the main challenges of contemporary science, but also of outstanding importance for the understanding and modeling of many real phenomena as public health (Colizza et al. 2007; Eubank et al. 2004; Vespignani 2009), information security management (Gross and Acquisti 2005) and the spread of ideas, opinions or innovations (Adamic and Glance 2005; Aral and Van Alsyne 2007; Díaz-Guilera et al. 2009; Domingos and Richardson 2001; Iribarren and Moro 2009). Since all these phenomena are driven by people, their dynamics is strictly related to the way in which individuals are connected and interact and to the mechanisms that govern the dynamics of such interactions. Every kind of social aggregation and interaction can be represented in terms of units composing this aggregation and relations between these units. One refers to this type of representation of a social structure with the term *social network* (Scott 2000; Wasserman and Faust 1994; Watts and Strogatz 1998), where sets of nodes are joined together in pairs by links or ties (Barabási 2003; Buchanan 2002; Strogatz 2001). In a social network every node represents a person, a group or organization, while relations between nodes are represented as a linkage between these units. Depending on the network and the context, the term relation may refer to acquaintance, kinship, group membership, scientific co-authorship or other. The advantage of such a representation is that it permits the analysis of social processes as a product of the nature and the relationships among social entities. In this context, the objects under observation are not only individuals and their attributes, but the relationships between individuals, their global structure, collective behavior and dynamics.

"The Convergence of Social and Technological Networks" (Kleinberg 2008).

G. Miritello, *Temporal Patterns of Communication in Social Networks*, Springer Theses, DOI: 10.1007/978-3-319-00110-4_1,

A quantitative analysis of the global properties of social networks started to develop in the early 1920s, focusing on characterizing the relationships among social entities as communication between members of a group, trades among nations, or economic transactions between corporations (Freeman 1996). Since then, an extensive research has been carried out to understand, measure and model the structure of many human-driven real world networks (Newman and Park 2003; Scott 2000; Wasserman and Faust 1994). The main outcome of this activity has been to reveal that real networks significantly deviate from random networks and that, despite their inherent differences, they are characterized by statistically similar topological properties (Albert et al. 1999; Newman 2003b; Strogatz 2001; Watts and Strogatz 1998). For instance, many real social networks have small characteristic path lengths, high clustering coefficients, are organized into groups or communities and are strongly heterogeneous: some individuals are more connected than others, the flux of information that pass through each social tie is unevenly distributed (Boccaletti et al. 2006; Díaz-Guilera 2007; Newman and Park 2003).

However, it was only in the last few years that the interest in social network science has grown so much as to radically change the scientific approach to model human behavior and society (Lazer et al. 2009). This interest has certainly been triggered by two main factors: the increasing availability of data with detailed information of human interactions on a large scale on one side and the optimized rating of computing facilities and optimization procedures on the other. Internet, mobile phones and online social networks have radically reshaped our way of communicating and interacting, giving us the possibility of being connected to everyone and everything all the time. By their nature, all these types of social interaction leave extensive digital traces of human habits. Mobile phone, e-mail and online instant messaging records document our social interactions, frequency and volume of communication; travel records and GPS navigation systems capture our physical locations and travel patterns; credit-card companies collect records of our buying habits, etc. Because of the incredible amount of sensitive information that such data contains, one of the most delicate challenges on data use is related to the privacy of users (Gross and Acquisti 2005). If properly anonymized, however, these data sets represent a huge scientific opportunity to precisely map human actions and develop models of social phenomena, allowing to unveil and explore, for the first time, the large scale characteristics of human behavior. Historically, collecting social-network data has been in fact a hard work and the scientific understanding of human society has been limited to relatively sparse observations of individuals or small groups and based on questionnaires reaching typically a few dozen individuals and relying on the individuals opinion to reveal the nature of human behavior. In contrast, a data-based analysis of human actions gives the opportunity to easily collect data on a very large scale and over extended periods of time. Obviously, with advantages come challenges. Connections that can seem trivial to analyze when only few individuals are involved can in fact become really hard and complex in larger networks. While small sets of data can be easily stored and processed, large scale data includes data sets with sizes (ranging from a few dozen terabytes to many petabytes in a single data set) beyond the ability of commonly-used software tools to capture, manage, and process. With this difficulty, new platforms of "big data" storage, statistics softwares and visualization packages have arisen in the few last years, changing the traditional database management tools.

The huge data capability, combined with the tools of network theory and statistics, has produced a dramatic advance in the understanding and modeling of many human-driven phenomena. Digital records can in fact be collected together into a general picture of individual and group behavior with the potential to transform the current understanding and modeling of humans and society. Dynamics and performance of groups and organizations can be analyzed through phone-call or e-mail data (Aral and Van Alsyne 2007; Eckmann et al. 2004; Guimerá et al. 2006; Wuchty and Uzzi 2011) and the analysis of the dynamics of such interactions can provide useful insights on how epidemic diseases or computer viruses spread through the population (Colizza et al. 2007; Lazer et al. 2009; Lloyd and May 2001). Traces of online communication such as Facebook or Twitter messaging offer another important source of data to understand peoples' interests, political and social opinions and sharing dynamics (Adamic and Glance 2005; Grabowicz et al. 2012; Watts 2007) offering new opportunities for research that would have been impossible otherwise.

Within this scenario, the approach to social science has been radically transformed to such an extent that the scientific community refers to the analysis of social networks as a new emerging type of science: the *network science* or *computational social science* (Barabási 2003; Giles 2012; Lazer et al. 2009; Watts 2004). At the same time, social science and social network analysis have become central tools for many organizational consultants and companies seeking to understand and model the connections between patterns of interactions and business outcomes such as job performance or satisfaction, adoption of new practices and technologies, information sharing, creation and propagation of new ideas/products/services. In addition, a quantitative understanding of information spreading, social influence processes or behavior change may certainly benefit many other areas, such as disease outbreak control, diffusion of ideas and innovations, public transport design, traffic engineering, and disaster management (González and Barabási 2007). This is the reason why the interest in analyzing social networks goes beyond the scientific interest.

All the social phenomena mentioned above are shaped by both the *topological structure* of the underlying network and the *temporal activity patterns* of human dynamics, since they depend not only on the way in which individuals are connected to each other and on the position they occupy within the network, but also on when and how they interact. Since both aspects are strictly correlated to one another, they can not be disentangled in the analysis of social networks. However, most of the current studies have focused on characterizing the topology and the formation processes of many social networks such as the web (Albert et al. 1999; Barabási et al. 2000; Faloutsos et al. 1999), transportation networks (Barrat et al. 2004; Vragović et al. 2005), scientific collaboration graphs (Barabási et al. 2002; Newman 2001b) or person-to-person communication networks (Ebel et al. 2002; Kossinets and Watts 2006; Kwak et al. 2010; Nanavati et al. 2008; Onnela et al. 2007), while the temporal dimension has been usually neglected. Traditional network models are therefore essentially static: once a snapshot of the contact network, i.e. who interacts with whom, is determined (by aggregating over time all the observed interaction events), it is assumed to be steady over time and any structural relationship between consecutive snapshots is ignored. However, real social networks are *temporal* networks (Holme and Saramäki 2012): they continuously change over time since agents appear and disappear (Palla et al. 2007; Tantipathananand et al. 2007), connections form, weaken, strengthen or decay in time (Burt 2000; Hidalgo and Rodriguez-Sickert 2008; Kossinets and Watts 2006), interaction events are time-ordered and governed by nontrivial dynamics and correlations (Onnela et al. 2003; Rybski et al. 2010; Saramäki et al. 2005; Zhao and Oliver 2010). Another assumption of static models is to suppose that interaction can be characterized by a weight or a rate of interaction events: thus, interaction events happen randomly in time and are not correlated to each other. As a consequence, human interaction can be approximated by homogeneous Poisson-like processes, which have been largely used to quantify many phenomena driven by humans (Haight 1967). One of the consequences of this assumption is that individual actions happen at relatively regular time intervals. In contrast, several empirical observations on humans-driven actions show that the human dynamics is instead high heterogeneous and *bursty* (Barabási 2010) and long periods of inactivities are followed by periods of intense bursts of activity (Barabási 2005; Karsai et al. 2011; Miritello et al. 2011; Vázquez et al. 2007). All these temporal inhomogeneities make human interactions significantly different from the Poissonian predictions (Iribarren and Moro 2009; Vázquez et al. 2007). Therefore, although the topology of a networks is indispensable to capture the underlying skeleton on which dynamical processes happen, it does not fully account for the complex temporal behavior of real processes. In the very last years much effort has been devoted to characterize the temporal properties of human interaction, understand their origin and quantify the impact they have on real phenomena such as information spreading or community formation (Barabási 2005; Hidalgo and Rodriguez-Sickert 2008; Iribarren and Moro 2009; Malmgren et al. 2008; Palla et al. 2007; Vázquez et al. 2007). However, a general understanding of how to incorporate the complex temporal patterns of contact dynamics into the current description of dynamical social networks is still lacking.

Within this challenging scenario lies the main motivation of this thesis: the expectancy that the understanding of the interplay between structure and dynamics would lead to a much better description and modeling of real social networks and their evolutionary mechanisms, as well as many real phenomena happening through the network. To address this, we treat and analyze different topics related to human behavior and social networks analysis, focusing on both the structure and the dynamics of large evolving communication networks. All the results and conclusions are supported by empirical observations and measurements of human interaction behavior conducted by analyzing massive longitudinal and cross-sectional (roughly 20 million people over a period of almost 2 years) anonymized databases of mobile phone calls traces. Because of the pervasive use of mobile phones, records of mobile communications collected by telecommunication companies represent a detailed and extended proxy of human interactions since they capture every phone call between any two parties of a large fraction of the population in a country, the initiation time and the duration of the call and localization in space of the parties.

The main goal of our analysis is twofold: provide a better understanding and characterization of human behavior and improve the current description and modeling of social networks as continuously changing entities, in contrast to the aggregated and static picture of the social structure. Our contribution is organized as follows. Chapter 2 contains a general overview of the common properties of topological and temporal patterns of real networks and how they are measured, with a special attention on the results present in the literature for social and communication networks. The objective of this chapter is to present basic concepts and preliminaries, introduce the notation and briefly survey the related work. We also discuss the general ideas of how social networks have been traditionally modeled, their main limitations and potential improvements. In Chap. 3 we address the problem of how people allocate time among their social connections. Since time and attention are inelastic resources, people only have a limited amount of time to dedicate to social interaction, thus they have to adopt a communication strategy to maintain their social circle. Some of the questions we address here are: do people actually follow any strategy of communication? How does such individual strategy relate to the individual and neighborhood properties? What are the global implications of the way in which people allocate their time and attention across their local network? Among others results, we show that the individual unbalance between the ability to establish new social connections and their capacity to maintain old ones gives rise to different strategies of social communication: while some people like to explore new part of the network, others instead keep very stable social circle. These communication strategies only emerge when taking into account not only the topological, but also the temporal patterns of human behavior and have important implications in global phenomena as network volatility and information diffusion. In Chap. 4 we analyze a fundamental and challenging topic in many social network and data mining tasks: the problem of predicting tie formation and decay. In social network analysis, this problem is strictly connected to the task of determining when an observed interaction between two individuals actually represents a social relationship. We first show how the current definition of a social tie can be largely improved by taking into account simple quantities that encompass the rhythm of tie communication, as the lifetime of the relation or the burstiness of the interaction events sequence between two individuals. Then, we propose a simple method which allows us to infer with high precision whether a tie is a newly-formed or a decayed social connection by only looking at its communication activity. The performance of our method is compared with the one of more standard models mainly based on topological aspects of the interaction network. Chapter 5 is devoted to understand how temporal patterns of communication affect dynamical phenomena and, in particular, spreading phenomena. Our main contribution to this issue is to give a quantitative analysis of how temporal aspects of human behavior, such as the bursty nature of interaction, the existence of group conversations or the tie creation/removal dynamics, affect the information spreading in social networks. We show that the interplay of these effects is crucial for both the reach and the speed of information in real networks, a result that have implications not only on the propagation of information between individuals, but also on many other spreading phenomena such as diffusion of innovations, social and political opinions and influence. We close the chapter by proposing a simple way to incorporate temporal patterns of human interactions into a static description of the network by means of what we have called the *dynamical strength* of social ties. The last chapter presents a summary and discussion of the main results of our work and its main implications in real phenomena. It also includes open questions and visions for future research.

An important part of the work of this thesis required the analysis of very large data sets of phone call communication. All the details about preparing, filtering and sampling the data are provided in Appendix A, together with further information about procedures and reference models used in the analysis.

References

Adamic L, Glance N (2005) The political blogosphere and the 2004 U.S. election: divided they blog. In: LinkKDD '05 Proceedings of the 3rd international workshop on link, discovery. ACM, New York, pp 36–43

Albert R, Jeong H, Barabási A-L (1999) The diameter of the world wide web. Nature 1999(491):130–131

Aral S, Van Alsyne M (2007) Network structure and information advantage. In: Proceeding of the academy of management conference, Philadelphia, PA

Barabási A-L (2003) Linked, the new science of networks: how everything is connected to everything else and what it means. Plume Books, New York

Barabási A-L (2010) Bursts: the hidden pattern behind everything we do. Dutton Books, New York

Barabási A-L, Albert R, Jeong H (2000) Scale-free characteristics of random networks: the topology of the world-wide web. Physica A 281:1–4

Barabási A-L, Jeong H, Ravasz R, Neda Z, Vicsek T, Schubert A (2002) Evolution of the social network of scientific collaborations. Phys A 311:590–614

Barabási A-L (2005) The origin of bursts and heavy tails in human dynamics. Nat Sci Rep 435:207–211

Barrat A, Barthélemy M, Pastor-Satorras R, Vespignani A (2004) The architecture of complex weighted networks. Proc Natl Acad Sci USA 101:3747

Boccaletti S, Latora V, Moreno Y, Chavez M, Hwang D-U (2006) Complex networks: structure and dynamics. Phys Rep 424:175

Buchanan M (2002) Small world: uncovering nature's hidden networks. Weidenfeld and Nicholson, London

Burt R (2000) Decay functions. Soc Netw 22:1–28

Colizza V, Barrat A, Barthélemy M, Vespignani A (2007) Predictability and epidemic pathways in global outbreaks of infectious diseases: the SARS case study. BMC Med 5:34

Díaz-Guilera A (2007) Complex networks: statics and dynamics. In: AIP conference proceedings, advanced summer school in physics 2006: frontiers in contemporary physics: EAV06, vol 885, pp 107–128

Díaz-Guilera A, Sergio L, Arenas A (2009) Propagation of innovation in complex patterns of interaction. In: Pyka A, Scharnhorst A (eds) Innovation networks: new approaches in modeling and analyzing, Springer, Berlin

Domingos P, Richardson M (2001) Mining the network value of customers. In: KDD '01 Proceedings of the 7th ACM SIGKDD international conference on knowledge discovery and data mining. ACM Press, New York, pp 57–66

Ebel H, Mielsch L, Bornholdt S (2002) Scale-free topology of e-mail networks. Phys Rev E 66:035103

Eckmann J-P, Moses E, Sergi D (2004) Entropy of dialogues creates coherent structures in e-mail traffic. Proc Natl Acad Sci USA 40:14333–14337
Eubank S, Guclu H (2004) Modeling disease outbreaks in realistic urban social networks. Nature 429:180–184
Faloutsos M, Faloutsos P, Faloutsos C (1999) On relationships of the internet topology. In: SIGCOMM computer communication, vol 29. ACM, New York, pp 251–262
Freeman LC (1996) Some antecedents of social network analysis. Connections 19(1):39–42
Giles J (2012) Computational social science: making the links. Nature 488:448
González M, Barabási A-L (2007) Complex networks: from data to models. Nat Phys 3:224–225
Grabowicz P, Ramasco J, Moro E, Pujol J, Eguiluz V (2012) Social features of online networks: the strength of intermediary ties in online social media. PLoS ONE 7(1):e29358
Gross R, Acquisti A (2005) Information revelation and privacy in online social networks. In: WPES '05 Proceedings of the 2005 ACM workshop on privacy in the electronic society, New York, pp 71–80
Guimerá R, Danon L, Díaz-Guilera A, Giralt F, Arenas A (2006) The real communication network behind the formal chart: community structure in organizations. J Econ Behav Organ 61:653–667
Haight FA (1967) Handbook of the poisson distribution. Wiley, New York
Hidalgo C, Rodriguez-Sickert C (2008) The dynamics of a mobile phone network. Phys A 387:3017
Holme P, Saramäki J (2012) Temporal networks. Phys Rep 519:97–125
Iribarren J, Moro E (2009) Impact of human activity patterns on the dynamics of information diffusion. Phys Rev Lett 103:038702
Karsai M, Kivelä M, Pan R, Kaski K, Kertész J, Barabási A-L (2011) Small but slow world: how network topology and burstiness slow down spreading. Phys Rev E 83:025102(R)
Kleinberg J (2008) The convergence of social and technological networks. Commun ACM 51(11):66–72
Kossinets G, Watts DJ (2006) Empirical analysis of an evolving social network. Science 311:5757
Kwak H, Lee C, Park H, Moon S (2010) What is Twitter, a social network or a news media? In: Proceedings of the 19th international conference on, world wide web. ACM, New York, pp 591–600
Lazer D, Pentland A, Adamic L, Aral S, Barabási A-L, Brewer N, Christakis NA, Contractor N, Fowler J, Gutmann M, Jebara T, King G, Macy M, Roy D, Van Alsyne M (2009) Computational social science. Science 323:721–723
Lloyd AL, May R (2001) How viruses spread among computers and people. Science 292:1316–1317
Malmgren RD, Stouffer DB, Motter AE, Amaral LAN (2008) A poissonian explanation for heavy tails in e-mail communication. Proc Natl Acad Sci USA 105:18153–18158
Miritello G, Moro E, Lara R (2011) Dynamical strength of social ties in information spreading. Phys Rev E 83:045102(R)
Nanavati AA, Singh R, Chakraborty D, Dasgupta K, Mukherjea S, Das G, Gurumurthy S, Joshi A (2008) Analyzing the structure and evolution of massive telecom graphs. IEEE Trans Knowl Data Eng 20:5
Newman MEJ (2001b) The structure of scientific collaboration networks. Proc Natl Acad Sci USA 98:404–409
Newman MEJ, Park J (2003) Why social networks are different from other types of networks. Phys Rev E 68:036122
Newman MEJ (2003b) The structure and function of complex networks. SIAM Rev 45:167–256
Onnela J, Chakraborti A, Kaski K, Kertész J, Kanto A (2003) Dynamics of market correlations: taxonomy and portfolio analysis. Phys Rev E 68:056110
Onnela J-P, Saramäki J, Hyvönen J, Szabó Z, Lazer D, Kaski K, Kertész J, Barabási A-L (2007) Structure and tie strengths in mobile communication networks. Proc Natl Acad Sci USA 104:7332
Palla G, Barabási A-L, Tamás V (2007) Community dynamics in social networks. Noise Stoch Complex Syst Finan 6601:660106
Palla G, Barabási A-L, Vicsek T (2007) Quantifying social group evolution. Nature 446:664–667

Rybski D, Buldyrev SV, Havlin S, Liljeros F, Makse HA (2010) Communication activity: temporal correlations, clustering, and, growth. arXiv:1002.0216v1
Saramäki J, Onnela J-P, Kertész J, Kaski K (2005) Characterizing motifs in weighted complex networks. In: Mendes JFF et al (eds) Science of complex networks, AIP conference proceeding, vol 776, pp 108
Scott J (2000) Social network analysis: a handbook. Sage Publications, London
Strogatz SH (2001) Exploring complex networks. Nature 410:268–276
Tantipathananand C, Berger-Wolf T, Kempe D (2007) A framework for community identification in dynamic social networks. In: KDD'07 Proceedings of the 13th ACM SIGKDD international conference on knowledge discovery and data mining, pp 717–726
Vázquez A, Rácz B, Lukács A, Barabási A-L (2007) Impact of non-poissonian activity patterns on spreading processes. Phys Rev Lett 98:158702
Vespignani A (2009) Predicting the behavior of techno-social system. Science 325:425–428
Vragović I, Louis E, Díaz-Guilera A (2005) Efficiency of informational transfer in regular and complex networks. Phys Rev E 71:036122
Wasserman S, Faust K (1994) Social networks analysis. Cambridge University Press, Cambridge
Watts DJ, Strogatz SH (1998) Collective dynamics of 'small-world' networks. Nature 393:440–442
Watts DJ (2004) The "new" science of networks. Annu Rev Sociol 30:243–270
Watts DJ (2007) Connections: a twenty-first century science. Nature 445:489
Wuchty S, Uzzi B (2011) Human communication dynamics in digital footsteps: a study of the agreement between self-reported ties and email networks. PLoS ONE 6(11):e26972
Zhao Q, Oliver N (2010) Communication motifs: a novel approach to characterize mobile communications. In: NetMob2010

Chapter 2
Social and Communication Networks

> *For all practical purposes, our behavior is random. Unpredictable. Episodic. Indeterminable. Unforeseeable. Irregular. There's only one problem with this assumption. It's simply wrong.*
>
> —Albert-Lázló Barabási.

As any progress in science, also the current description of social networks has been accomplished one step at a time. Traditionally, the study of social and complex networks has been the territory of graph theory, which allows to define a network for any system by means of the simple representation of a graph. Since the 1950's real networks have been described as completely random graphs (Bollobás 1985), proposed as the simplest description of connections between entities and which mathematical formulation inheres in the work of Paul Erdös and Alfréd Rény (Erdös and Rényi 1959, 1960). According to this formulation, any member of the network has the same probability to be connected to any one else, thus all members have approximately the same number of connections. This vision radically changed in the late '90's with the empirical evidence that actually the structure of real complex systems is far from random (Barabási and Albert 1999; Watts and Strogatz 1998). In fact, in many real social networks, individuals exhibit preferential connectivity (higher probability that one of them will be linked to another one that already has a large number of connections), some connections are much more strong (or important) than others, social interactions are organized in tight groups or communities (Barabási and Albert 1999; Granovetter 1973; Watts and Strogatz 1998). Due to the increasing evidences of these structural heterogeneities, a big effort has been done from the scientific community to measure the topological properties and to understand, model and predict the mechanisms which regulate the formation of such structure and their impact on real phenomena. The standard way to address this has been focused on determining the contact network (who interacts with whom) in given time window, then characterize

"Bursts: The Hidden Pattern Behind Everything We Do" (Barabási 2010).

G. Miritello, *Temporal Patterns of Communication in Social Networks*,
Springer Theses, DOI: 10.1007/978-3-319-00110-4_2,

its elements by the aggregated properties measured in that time window and try to model the network dynamics to explain the observed structure. According to this frame, a time-aggregated and static picture of the network is given, where any inherent dynamics characterizing human behavior is neglected: the aggregated volume of interaction or the flux passing through a given link is the main quantity to assess the importance of that connection thus it fully determines its strength (or importance), interactions can happen at any time, there is no causality between events, communication is homogeneous in time, etc. This traditional approach, to which we will refer as *static approach*, was motivated by the expectancy that the characterization of the network structure would lead to a better knowledge of its dynamical and functional behavior. However, real social networks are dynamical objects, whose interactions between their members happen at a given time, may have a given duration and a causal relation. In this respect, the assumption of projecting out the temporal dimension may discard important information about the dynamical properties of real networks, their correlations with the topological ones and the dynamics of real phenomena. Only in the very last years, the large availability of massive databases of human behavior and interaction patterns such as e-mail, phone calls or online interactions databases led to the observation of non-trivial temporal properties of real social networks, with important implications on the way in which social networks and real phenomena have been traditionally understood and modeled. It has been observed, for example, that contrary to the predictions of statics approaches, individual actions do not happen at any time nor with the same probability to any other member of the network, that social interactions are not everlasting but, in contrast, they appear to be very unstable and volatile (Barabási 2005; Eckmann et al. 2004; Hidalgo and Rodriguez-Sickert 2008; Kossinets and Watts 2006; Vázquez et al. 2006). These results indicate that, as well as the (static) connections between members of a network, not even the temporal patterns of human actions and interactions can be modeled as random. Once again, therefore, the vision of how to model real networks is experiencing a radical change.

Without claiming to be complete, this chapter is intended to serve as a brief introduction of what social networks are, how they can be characterized and what are their main properties and features known in the literature. The first part of the chapter is dedicated to those topological properties of social networks that will be referenced in the rest of the thesis, what they represent and how they have been traditionally measured. Then we get more into the particular case of social networks that constitutes the main subject of this thesis: communication networks. We concentrate mainly on those aspects of communication networks that make them different from other social networks. After presenting the basic guideline of how the topology of social networks have been historically modeled, we discuss the main limitation of such traditional approaches on the basis of those temporal aspects of social networks that have been observed and measured only in the very last years. The latter part offers a first overview of the crucial role played by temporal aspects of human interaction into the characterization of social networks. At the same time it constitutes the starting point of all the work done in this thesis, presented in the following chapters.

2.1 Topology of Social Networks

Quantifying the topology is the first step towards detecting patterns in the network. The understanding of how a graph is wired, whether its member are equally connected or not or if every zone of the graph is reachable from any other one, gives very important insights on the network and the dynamical system it represents. For example, the way in which a disease or a piece of information would spread across society or how opinions form depends, above all, on the topology of the underlying graph. Nevertheless in many cases what we observe of a network is not the structure itself, but several instantaneous interactions between agents. Recover the very structure of how individuals are wired can be, therefore, a hard task. In most network studies the solution has been to aggregate all the interactions observed in the whole observation time window between any two members into a static edge between them, and possibly use the total number of interactions to asses the intensity or importance of the connection. Although, as we will see in the rest of this thesis, it represents only the first approximation of a network, this approach allowed to define an abundance of useful quantities to measure the main properties of real networks (Costa et al. 2007) and led to the discovery of several features which make them peculiar with respect to other types of networks, such as technological and biological networks (Newman and Park 2003).

2.1.1 Definitions and Notations

A network is a set of items, called vertices (or nodes), with connections between them, called edges (or links or ties) (Fig. 2.1). In mathematical terms a network is represented by a *graph* (West 1995). A undirected (directed) graph is a pair of sets $\mathcal{G} = \{\mathcal{P}, \mathcal{E}\}$, where $\mathcal{P}$ is a set of $\mathcal{N}$ nodes (or vertices) $p_1, p_2, ..p_N$ and $\mathcal{E}$ is a set of $\mathcal{M}$ edges (or links) that connect two elements of $\mathcal{P}$. A vertex typically represents an object (or an individual), while an edge represents a relation between two objects (or the same object). In a undirected graph, as the one depicted in Fig. 2.1a, each of the links is defined by a couple of nodes i and j, and is denoted as (i, j), e_{ij} or simply ij. In a directed graph, the order of the two nodes is important: e_{ij} stands for a link from i to j, and $e_{ij} \neq e_{ji}$. The property that two nodes in a directed network point to each other is called *reciprocity*. For a graph $\mathcal{G}$ with $\mathcal{N}$ nodes, the number of edges $\mathcal{M}$ is at least 0 and at most $\mathcal{N}(\mathcal{N} - 1)/2$ (when all the nodes are pairwise adjacent). Graphs are usually represented as a set of dots, corresponding to the nodes, which are joined together by a segment if the corresponding nodes are connected by a link (see Fig. 2.1). Usually, it may be useful to consider a matricial representation of a graph. Given a graph $\mathcal{G} = \{\mathcal{P}, \mathcal{E}\}$, it can be in fact completely described by giving the adjacency matrix $\mathcal{A}$, which is a $\mathcal{N} \times \mathcal{N}$ matrix whose element a_{ij} $(i, j = 1, ..., n)$ is equal to 1 if an edge e_{ij} exists and 0 otherwise. For undirected graphs $\mathcal{A}$ is thus

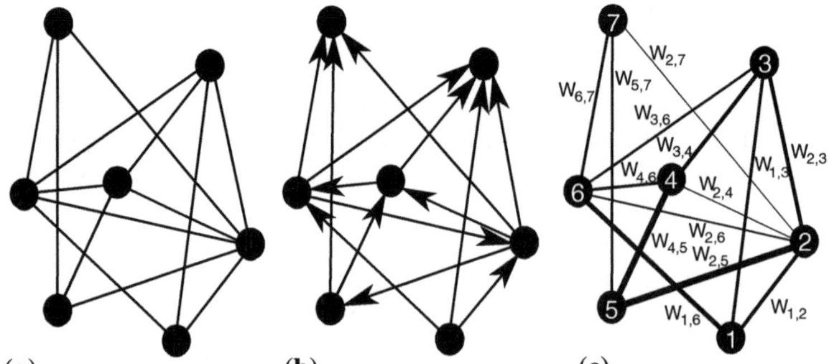

(a) (b) (c)

Fig. 2.1 Schematic representation of **a** a undirected, **b** a directed and **c** a weighted (and undirected) graph with N = 7 nodes and M = 14 edges. Adapted from "Complex networks: Structure and dynamics", Boccaletti et al. (2006)

a symmetric matrix. A *subgraph* $\mathcal{G}' = \{\mathcal{P}', \mathcal{E}'\}$ of the graph $\mathcal{G} = \{\mathcal{P}, \mathcal{E}\}$ is a graph with $\mathcal{P}' \subseteq \mathcal{P}$ and $\mathcal{E}' \subseteq \mathcal{E}$.
Although two nodes are not adjacent, they may however be reachable from one to the other. An important concept in graph theory is in fact the concept of *walk* or *path* from node i to node j, defined as a sequence of nodes and edges that begins with i and ends with j. The length of the walk is defined as the number of edges in the sequence.

In the case of social networks nodes represent a set of individuals or social entities linked through some kind of social interactions among them such as friendship, kinship, status, sexual, business or political, which define the links among them (Scott 2000; Wasserman and Faust 1994). Due to the recent development of communication systems, such as Internet or mobile phones, many other examples of social networks can be actually defined (Wellman et al. 1996; Wellman 2001), such as networks of human interaction through e-mail, Web forms, mobile phone or online social networks and services as Facebook, LinkedIn, Twitter. We refer to the latter as *communication* or *interaction* networks.

Node Degree and Assortative Mixing

The *degree* or *connectivity* k_i of a node i is the number of edges it has to other nodes and can be defined in terms of the adjacency matrix $\mathcal{A}$ as:

$$k_i = \sum_{j \in \mathcal{N}} a_{ij}. \tag{2.1}$$

In the case of a directed graph, the degree of a node is defined as the sum of the *out-degree* $k_i^{out} = \sum_j a_{ij}$ and the *in-degree* $k_i^{in} = \sum_j a_{ji}$, which measure respectively the number of outgoing and ingoing edges. Connectivity is a fundamental concept for networks. In real networks, not all nodes have the same number of edges and a first characterization of the network can be indeed obtained by dividing it in groups of nodes according to their connectivity. Especially for large networks, a more convenient characterization is obtained in terms of a distribution function (*degree distribution*) $P(k)$, which gives the probability that a randomly selected node has k edges. For directed networks, both $P(k^{in})$ and $P(k^{out})$ are defined. The n-moment of $P(k)$ is given by:

$$\langle k^n \rangle = \sum_k k^n P(k). \tag{2.2}$$

The first moment $\langle k \rangle$ defines the mean degree of the graph $\mathcal{G}$ and the second moment $\langle k^2 \rangle$ measures the fluctuations of the degree distribution. As mentioned above, until the late '90's many real networks have been modeled as random graphs (Erdös and Rényi 1959). A key prediction of random network theory is that most nodes have approximately the same degree, close to the average $\langle k \rangle$ of the network. In this case the degree distribution is a bell-shaped Poisson distribution with a peak at $P(\langle k \rangle)$, as the one depicted in Fig. 2.2a. Finding nodes that have a significantly greater or smaller number of links than a randomly chosen node is therefore rare. One also refers to random networks as exponential networks since the probability that a node is connected to other k nodes decreases exponentially (Haight 1967). For many real networks however, it has been found that $P(k)$ displays a power law shaped degree distribution $P(k) \sim k^{-\gamma}$ (see Fig. 2.2b), with exponent varying in the range $2 < \gamma < 3$ (Albert et al. 1999; Faloutsos et al. 1999; Jeong et al. 2000). In these networks, the average degree $\langle k \rangle$ is therefore well defined and bounded, while the variance $\sigma^2 = \langle k^2 \rangle - \langle k \rangle^2$ is dominated by the second moment of the distribution, which is highly fluctuating. Contrary to random networks, the average degree is not anymore a meaningful characterization of the network properties. Due to the property of power-laws of having the same functional form at all scales, such networks are referred as *scale-free networks* (Barabási and Albert 1999) and have been the focus of a great deal of attention in the literature (Albert and Barabási 2002; Dorogovtsev and Mendes 2002; Strogatz 2001). For many real networks, actually, $P(k)$ displays an exponential cutoff. However, its functional form still deviates significantly from the Poisson distribution expected for a random graph. Indeed, contrary to random networks, scale-free networks have a highly heterogeneous degree distribution, which results in the simultaneous presence of a few nodes (also called *hubs*) linked to many other nodes, and a large number of poorly connected elements.

The most known model to explain the origin of this scale invariance is the *Albert-Barabási* model (Barabási and Albert 1999) which is based on two basic ingredients: growth and preferential attachment, two key features of real networks. According to this model, a node with more links increases its connectivity faster than nodes with fewer links, since incoming nodes tend to connect to it with higher probability, a mechanisms that actually leads to the appearance of an hub hierarchy that exemplifies the scale-free structure (Bollobás et al. 2001; Dorogovtsev et al. 2000).

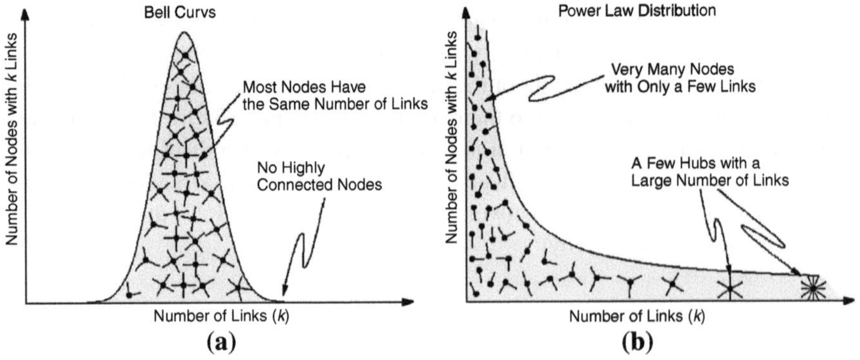

Fig. 2.2 Degree distribution $P(k)$ for a **a** random and **b** scale-free network. For random networks $P(k)$ follows a *bell-shaped* Poisson distribution where most nodes have the same number of connections. In contrast, in scale free-networks the degree distribution is power-law shaped and indicates that most nodes have only a few connections, while a few nodes are very highly connected (hubs). Adapted from "The architecture of complexity", Barabási (2007)

The *Albert-Barabási* model has attracted an exceptional amount of attention in the literature. In addition to analytic and numerical studies of the model itself, many authors have proposed modifications and generalizations to make the model a more realistic representation of real networks such as models with nonlinear preferential attachment, dynamic edge rewiring, fitness models and hierarchically and deterministically growing models (Albert and Barabási 2002; Dorogovtesev and Mendes 2001; Huberman and Adamic 1999; Goh et al. 2002; Gómez-Gardeñes and Moreno 2004).

Another class of models is based on three mechanisms: duplication (a randomly selected node and all its connections are duplicated); divergence (connections of a duplicated node are re-moved with a give probability) and mutation (connections are added from the duplicated node to a fraction of nodes which are not neighbors of the original node). The latter model also produces power-law degree distribution, although it fails to predict other properties of real networks that we will introduce in the rest of the chapter, such as degree correlations and the clustering coefficient (Solé et al. 2002). A better prediction for clustering coefficient is instead given by models that only includes duplication and divergence (Vázquez et al. 2003).

The fat tail of the degree distribution and the divergence of the second moment can affect the properties of a network, such as the clustering coefficient (Newman and Park 2003), which in many real social networks is much higher than the one expected for the corresponding random model (Amaral et al. 2000; Newman et al. 2002; Watts and Strogatz 1998). The heterogeneity of scale-free connectivity patterns also affects the behavior of dynamical processes that take place over the graph such as spreading processes. It has been shown, for example, that the presence of large degrees nodes favors epidemic spreading not only by suppressing the epidemic threshold, but also by accelerating the virus propagation in the population (Barthélemy et al. 2004; Moreno et al. 2002; Pastor-Satorras and Vespignani 2001b), a topic that will be analyzed in Chap. 5. The implications of this result can be very important in the set-up of dynamic control strategies, such as targeted immunization strategies, in populations with heterogeneous connectivity patterns (Pastor-Satorras and Vespignani 2002).

The degree distribution $P(k)$ describes all the statistical properties of uncorrelated networks. However, in many real networks the probability that a node with degree k is connected to a node with degree k', depends on k (correlated networks). In these cases, it is convenient to define the conditional probability $P(k'|k)$ which represents the probability that a node with degree k is connected to a node with degree k'. Another measure of degree-degree correlation is the *average nearest neighbors degree* of a node i:

$$k_{nn,i} = \frac{1}{k_i} \sum_{j \in N_i} k_j = \frac{1}{k_i} \sum_{j=1}^{N} a_{ij} k_j, \tag{2.3}$$

where N_i is the set of neighbors of node i. By means of this definition one can calculate the average degree of the nearest neighbors of nodes with degree k. The latter quantity, denoted as $k_{nn}(k)$, is related to the conditional probability as $k_{nn}(k) = \sum_{k'} k' P(k'|k)$ and implicitly incorporates the dependence on k. In absence of degree correlations one gets $k_{nn} = \langle k^2 \rangle / \langle k \rangle$, which shows the independence of k_{nn} on k. For correlated graphs, instead, k_{nn} depends on k. Depending on whether k_{nn} is an increasing or decreasing function of k, this properties is known as *assortative* or *disassortative* mixing (Newman 2002a). While in assortative networks nodes tend to connect with nodes with similar degree, in disassortative networks nodes with high degree are more likely connected with lowly connected ones.

Degree correlations and assortative mixing are very important properties in social networks since they have implications for questions as diffusion of information, network resilience or vaccination strategies (Callaway et al. 2000; Pastor-Satorras and Vespignani 2002). In disassortative networks, for example, a path between pairs of vertices can be destroyed by the removal of just a few of the highest degree nodes. Attacks on the highest degree vertices are therefore much more effective since the removal of few of them leads to a fast collapse of the whole network. On the contrary, in assortative networks, the removal of high-degree nodes is a relatively inefficient strategy for destroying network connectivity, since these vertices tend to be clustered together, thus their removal would result to be ineffective (Newman 2002a). In Sect. 2.2.1 we will see that in many social networks assortativity emerges as a natural phenomenon not only in the degree, but also with respect to a variety of psychological, sociodemographic and behavioral attributes (Christakis and Fowler 2007; Lewis et al. 2008; McPherson et al. 2001).

Shortest Path Length, Diameter and Betweenness

The shortest path d_{ij} between two nodes i and j measures the geodesic or optimal path way that go from i to j. A measure of the typical separation between two nodes in a graph $\mathcal{G}$ is given by the *average shortest path length* l, defined as the mean geodesic distance between nodes pairs (Watts 1999):

$$l = \frac{1}{N(N-1)\sum_{i,j \in N, i \neq j}} d_{ij}. \tag{2.4}$$

The maximum value of d_{ij} is called the *diameter* of the graph. In networks with more than one component (maximally connected induced subgraph), the definition in (2.4) can be problematic since there exist nodes pairs that have no connecting path. One can assign infinite geodesic distance to such pairs, but then the value of l also becomes infinite. For this reason, on such networks one usually defines l as the mean geodesic distance between all pairs of nodes belonging to the largest connected component (Latora and Marchiori 2001). Despite their large size, most of the real networks usually show a relatively short path between any two nodes. This feature is known as the *small-world effect* and is mathematically characterized by l, that depends at most logarithmically on the network size N (Watts and Strogatz 1998; Watts 1999).

In the social context, the small-world effect was first investigated by Milgram in the 1960s, in a series of experiments to estimate the actual number of steps in a chain of acquaintances (Milgram 1967). In its first experiment, Milgram asked randomly selected people in Nebraska to send letters to a distant target individual in Boston, identified only by his name, occupation and rough location. The letters could only be sent to someone whom the current holder knew by first name, and who was presumably closer to the final recipient. Milgram kept track of the paths followed by the letters and of the demographic characteristics of their handlers. Although the common guess was that it might take hundreds of these steps for the letters to reach their final destination, for those letters which arrived at destination, Milgram found that it had only taken an average of six steps for a letter to get from Nebraska to Boston. He labeled this situation "six degrees of separation" (Guare 1990), a phrase which since then has passed into popular folklore. Although the experiment certainly contained many possible sources of error, the general result that two randomly chosen persons can be connected by a short chain of intermediate acquaintances has been subsequently verified, and it is now widely accepted (Dodds et al. 2003; Korte and Milgram 1970).

Two nodes i and k that are not directly connected by an edge in a graph, can be linked through the nodes belonging to all the paths connecting i and k. In this regard, a measure of the relevance of a given node is given by its *betweenness* b_i which measures the number of geodesics going through it and is defined as:

$$b_i = \sum_{j,k \in N, j \neq k} \frac{n_{ij}(i)}{n_{jk}}, \tag{2.5}$$

where n_{jk} is the number of shortest paths connecting j and k, while $n_{jk}(i)$ is the number of shortest paths connecting j and k passing through i. Together with the degree, the betweenness of a node is one of the standard measures of the centrality of a node in a network (Scott 2000). Betweenness centrality can also be viewed as a measure of network resilience, indicating how much effect on path length the removal of a vertex will have (Holme et al. 2002; Newman 2003b).

Clustering or Transitivity

In many networks it is found that if node i is connected to node j and node j to node k, then there is a high probability that node i will also be connected to node k. In the language of social relationships, this translate in: the friend of your friend is likely also to be your friend (Wasserman and Faust 1994). In the study of networks, this property is known as *transitivity* or *clustering* and, in terms of a generic graph $\mathcal{G}$ it means the presence of a high number of triangles (sets of three vertices each of which is connected to each of the others). This can be quantified by defining the clustering coefficient $\mathcal{C}$ thus:

$$\mathcal{C} = \frac{3 \times \textit{number of triangles in the graph}}{\textit{number of connected triples of vertices in the graph}}, \quad (2.6)$$

where a triple consists of three nodes connected by two (open triple) or three (close triple) and the factor 3 in the numerator accounts for the fact that each triangle contributes to three triples and ensures that $\mathcal{C}$ lies in the range $0 \leq \mathcal{C} \leq \infty$. An alternative definition of the clustering coefficient has been given by Watts and Strogatz (Watts and Strogatz 1998). It is obtained by defining a *local clustering coefficient* of a node i as

$$\mathcal{C}_i = \frac{\textit{number of triangles connected to node i}}{\textit{number of connected triples centered on node i}}. \quad (2.7)$$

One assumes $\mathcal{C}_i = 0$ for nodes with degree 0 or 1, for which both numerator and denominator are zero. The clustering coefficient for the whole network is given by the average

$$\mathcal{C} = \frac{1}{N} \sum_i \mathcal{C}_i, \quad (2.8)$$

and, by definition, $0 \leq \mathcal{C} \leq 1$. In sociologic literature, the local clustering coefficient $\mathcal{C}_i$ has been widely used as a measure of the "network density" (Scott 2000). In general for social networks, regardless of which definition of the clustering coefficient is used, the values of $\mathcal{C}$ tend to be considerably higher than for a random graph with a similar number of vertices and edges. This indicates that nodes tend to create tightly connected groups characterized by a relatively high density of ties (Watts and Strogatz 1998). According to Newman and Park, together with the degree correlations, this property is what makes social networks different from other networks (Newman and Park 2003).

Communities

Given a graph $\mathcal{G}$, a *community* is a cohesive subgraph $\mathcal{G}'$ whose nodes are tightly connected. Since the structural cohesion of the nodes of a graph can be quantified in several different ways, there are many formal definitions of community structures (Ahuja et al. 1993; Everitt 1974; Girvan and Newman 2002; Guimerá et al. 2003; Holme 2002; Newman and Girvan 2004; Wilkinson and Huberman 2004). The most known definition is based on the concept of a *clique* and requires that all pairs of community members have relationships with each other. A clique is a maximal complete subgraph of three or more nodes, that is a subset of nodes all of which are adjacent to each other, and such that no other nodes exist adjacent to all of them. By extending this definition, a *n-clique* is a maximal subgraph in which the largest geodesic distance between any two nodes is no greater than n. Community structures are a typical feature of social networks; it is a matter of common experience that people divide into groups along lines of interest, occupation, location, age or family ties (Newman and Girvan 2004). The different way in which an individual is embedded in the structure within the network is also related to the behavior or function he is likely to practice. Individuals belonging to tightly connected group may be crucial in providing emotional and material support to each other (Granovetter 1973; Wellman 2007), while individuals who act as bridges between groups may provide access to a greater variety of information (Eagle et al. 2010; Onnela et al. 2007b).

Social analysts were the first to formalize the idea of communities and develop mathematical measures and methods to define the cohesion of communities and identify subgroups (Wasserman and Faust 1994; Scott 2000). In the last few years there has been an increasing interest and research in this area, which has become one of the most prominent areas of network science (Arenas et al. 2010; Blondel et al. 2008; Danon et al. 2008; Kumpula et al. 2009; Lancichinetti et al. 2010a,b; McDaid 2010; Newman and Girvan 2004; Raghavan et al. 2007; Rosvall and Bergstrom 2008; Toivonen et al. 2006, 2007). Finding the communities within a network is in fact a powerful tool for understanding not only the structure and the growth mechanisms of a network, but also its functioning: a community in a social network might indicate a circle of friends, a community in the World Wide Web might indicate a group of pages on closely related topics, and a community in a cellular or genetic network might be related to a functional module.

Motifs

A *motif* $\mathcal{M}$ in a network is a pattern of interconnections occurring in a graph $\mathcal{G}$ at a number significantly higher than in randomized versions of the graph, i.e. in graphs with the same number of nodes, links and degree distribution as the original one, but where the links are distributed at random. $\mathcal{M}$ can be considered as a sub-graph of $\mathcal{G}$. The concept of motif was first studied for biological networks (Milo et al. 2002; Mangan and Alon 2003) and then extended to other networks from neurobiology and ecology to social networks (Zhao and Oliver 2010). The reasons of the high frequency of different subgraphs in a specific network are not totally understood. There are at least two possible explanations. On the one hand, it is possible that certain constraints on the growth mechanism of a network as a whole determine which motifs become abundant. On the other hand, it is well known that the structure has important consequences on the network dynamics and functional robustness (Valverde and Solé 2005; Vázquez et al. 2004). So that a particular sub-graph can become overrepresented because, due to its structure, it possesses some relevant functional properties (Milo et al. 2002). As we will see in Sect. 2.4.3 the presence of motifs can also reveal important correlations between the agents, which may have a causal explanation.

2.1.2 Weighted Networks

Up to now, we have focused on networks in which edges between nodes have a binary nature, in the sense that they are either present or not. This networks are known as unweighted networks and each connection is assumed to be equivalent to any other. Nevertheless, in many real networks not all edges have the same importance of role and display instead a large heterogeneity in the capacity and the intensity of the connections. For example, social relationships with family members are usually different from friends or acquaintances (Granovetter 1973; Roberts 2010; Wellman and Wortley 1990). In all these cases it may be useful to assign to edges attributes that somehow should allow to distinguish connections of different type. These systems are better described in terms of weighted networks, i.e. networks in which each link carries a numerical value w_{ij} measuring the *strength* or *weight* of the connection (see Fig. 2.1c). Depending on their weight, ties can be distinguished between "strong" (large weight) and "weak" ties (small weight). Examples of strong and weak ties are found in social networks (Barabási et al. 2002; Granovetter 1973; Latora and Marchiori 2001; Newman 2001a) as well as other types of networks such as neural networks (Latora and Marchiori 2001, 2003; Sporns 2003), airline networks (Barrat et al. 2004; Guimerá et al. 2005), network of pages on the Internet (Pastor-Satorras and Vespignani 2004), biological systems (Csermely 2004). In most of these networks, tie strength quantifies the attention or the flow of information through that connection. As we will see in the next section, in the case of social networks such as mobile phone, e-mail or online social networks, the weight of a tie is usually assigned depending on the volume of interaction between the two involved individuals in a given time window, which has been found to correlate with the intensity of the social relationship (Baym et al. 2004; Wellman and Haythornthwaite 2003). Exploring the strength of the ties in social networks helps in the understanding of the structure of the network and also of the dynamics of many phenomena that involve human behavior, such as communities formation, information spreading and social influence (Grabowicz et al. 2012; Onnela et al. 2007b; Toivonen et al. 2007; Watts 2004). To measure the level of edge reciprocity in weighted (directed) networks, one can also define the bias $b_{ij} = w_{ij}/(w_{ij} + w_{ji})$ (which takes values from 0 to 1), where w_{ij} and w_{ji} are respectively the number of calls from i to j and from j to i. Several generalizations of the quantities defined in the previous section for unweighted networks (average nearest neighbors degree, the shortest path length, the clustering coefficient or motifs) have been proposed to characterize the complex statistical properties and heterogeneity of weighted social networks (Barrat et al. 2004; Boccaletti et al. 2006; Onnela et al. 2005; Saramäki et al. 2007).

Node Strength and Strength Distributions

In a weighted graph each edge has a weight w_{ij}, which is equal to 0 if the nodes i and j are not connected. It has been found that the weights characterizing the various connections exhibit generally complex statistical features with highly varying distributions and power-law behaviors (Almaas et al. 2004; Colizza et al. 2006; Goh et al. 2001; Onnela et al. 2007b). For a node i, the sum of w_{ij} over all his connections defines the node strength s_i:

$$s_i = \sum_j w_{ij}, \tag{2.9}$$

which is a measure of node strength in terms of the total weight of its connections (Barrat et al. 2004; Yook et al. 2001; Onnela et al. 2003). When the weights are independent on the topology, the strength of nodes of degree k is $s(k) \sim \langle w \rangle\, k$, where $\langle w \rangle$ is the average tie weight across the whole network. In the presence of correlations, instead, one has $s(k) \simeq A\, k^{\beta}$ with $\beta \neq 1$ and $A \neq \langle w \rangle$ (Barrat et al. 2004; Miritello et al. 2012b). Together with the degree distribution $P(k)$, the strength distribution $P(s)$, which measures the probability that a node have strength s, gives useful information about the network. In many real network the node strength is related to the node degree, thus $P(s)$ is also heavy-tailed. The same is observed for the distribution of tie weights. The local coupling between node and tie strength and network topology has important consequences for the network's global stability if ties are removed, as well as for the spread of news and ideas within a network (Onnela et al. 2007b).

Node Disparity

For a node i with a given connectivity k_i and a given strength s_i, there are different combinations of w_{ij}. In the two limit cases, all weights can be of the same order s_i/k_i or, in contrast, only one or few weights can dominate over the others. To measure this level of diversity, a quantity which is widely used in network literature is the *disparity* Y_i (Almaas et al. 2004; Barthélemy et al. 2003, 2005; Boccaletti et al. 2006; Miritello et al. 2012b), given by:

$$Y_i = \sum_{j=1}^{k_i} \left(\frac{w_{ij}}{s_i} \right)^2. \tag{2.10}$$

The disparity is a measure of local heterogeneity and it has an implicit dependence on k_i. In the homogeneous case, in which all edges have comparable weights, $Y_i(k) \sim 1/k$ since $w_{ij} = s_i/k_i$. In contrast, if the weight of a single edge dominates, then $Y(k) \simeq 1$ and it is independent of k. Other measures to quantify the topological diversity in a network have also been used, as the Shannon Entropy H_i or the Rényi Disparity $D_i(\gamma)$, where γ is a tunable constant (Eagle et al. 2009; Lee et al. 2010). These quantities however are strongly related to each other: in fact H_i behaves like $1/Y_i$ and D_i reduces to $1/Y_i$ in the case $\gamma = 2$, while for $\gamma = 1$ it reduces to the Shannon disparity, which is the exponential of the Shannon entropy.

2.2 Communication Networks

Communication (or interaction) networks, as the same word indicates, are a particular case of social networks where interactions between individuals refer to sequences of communication events. Examples of communication networks include e-mail networks, text messages or phone-calls, communication via blogs, online friendship networks as Facebook or micro-blogging services as Twitter. The increasing understanding and modeling of many real social networks can actually be attributed to the analysis of communication networks derived from massive electronic data set generated by millions of people through all these communication channels (Lazer et al. 2009). These records, which are routinely collected on websites, communication companies or electronic providers, represent a very rich laboratory to study how humans act and interact and to understand phenomena such as computer viruses or disease spreading (Anderson and May 1992; Newman 2002b; Newman et al. 2002), diffusion of information, innovations and products (Aral and Van Alsyne 2007; Dodds and Watts 2003; Szabó and Barabási 2006), opinion and influence dynamics (Friedkin and Johnsen 1990; Quattrocchi et al. 2010), teams formation (Guimerá et al. 2005) and so on. In contrast to others social networks, edges in communication networks typically arise from instant communication events and capture relationships as they happen. At any given instant, in fact, the network consists of the collection of ties connecting the people who are currently having a conversation. Together with the large dimension of these databases, this is certainly one of the main advantages of the study of communication networks. However, while in networks such as co-authorship networks a relationship between two scientists can be easily inferred whenever they coauthored at least one paper together (Newman 2001a), in many communication networks the problem of tie definition is usually not so trivial. For example, in mobile phone networks not all the observed events necessarily correspond to a social relationship between the individuals involved, since they may refer to calls to the operator or to wrong numbers. At the same way, declared "friends" in

Facebook may not correspond to the real individual social circle, since users actually interact only with few of them (Baym et al. 2004; Feld 1991). In this respect, to infer unobserved social relationships from the observed communication events is usually a hard task (De Choudhury et al. 2010; Wuchty 2009). Another important thing to bear in mind when dealing with networks of human communication is the different nature they may have. Although it is not always relevant, sometimes it can be important to distinguish between *offline* and *online* networks, where *offline* usually refer to phone calls or short messages (SMS) networks, to distinguish them from *online* communication such as blog interaction, Facebook, LinkedIn or Twitter networks. Another possible classification is related to the number of individuals involved at each interaction event. In this respect, while mobile phone or text services networks typically represent *one-to-one* communication networks, in the case of e-mail, Twitter or Facebook one may talk about *one-to-many* communication networks, since interaction can involve many recipients. Finally, it is often important to account for the directional/undirectional nature of the communication. While a phone call allows a bidirectional communication and the party that initiates the call may be only partially relevant, when dealing with SMS, e-mails or direct messages, to distinguish between the sender and the receiver may be instead crucial to characterize the underlying social relationship.

Each of these communication channels represents of course only one of the several possible ways in which two individuals may be connected in the real life. However, the use of electronic data as a proxy for social interactions has already proved successful in several recent investigations. For example, in the case of mobile phone networks, it has been observed that communication ties constitute an accurate representation of face-to-face interaction and self-reported friendships as measured using traditional sociometric methods (Eagle et al. 2009). This allows the quantification of previously rather elusive quantities such as tie strength that serve as signatures of work, family or acquaintances relationships (Onnela et al. 2007a,b), led to analyze how social groups evolve and change over time (Palla et al. 2007a,b). Other studies of e-mail communication networks have also shown that the use of e-mail in local social circles is strongly correlated with face-to-face and telephone interactions (Baym et al. 2004; Wellman and Haythornthwaite 2003; Wuchty and Uzzi 2011) and that the patterns of e-mail communication are related to the underlying social structure, shared activities, and personal attributes (Kossinets and Watts 2006; Wuchty and Uzzi 2011). Questions like how well electronic communication represent real social relationships or until what extent social media can predict the intimacy of a social relationships have been addressed also in online settings (Adamic and Adar 2003; Golder et al. 2007; Kivran-Swaine et al. 2011; Ugander et al. 2011). For example, a comparison between a network of Facebook interactions and self-reported data revealed that it is possible to infer the existence and the importance of a *offline* social contact by looking at the *online* network of the involved individuals and its properties such as the number of exchanged messages, number of common friends, etc. (Gilbert and Karahalios 2009). Furthermore, location-based networks such as Foursquare or GPS records of mobile phone networks provide important information on the physical places that people visit and how they move (Scellato et al. 2011; Volkovich et al. 2012), by providing more interesting insights about human mobility patterns (Candia et al. 2008; González et al. 2008) or the relations between friendships and mobility (Cho et al. 2011; Cranshaw et al. 2010).

However, despite the success of the use of electronic communication to represent and study social relationships, in social network analysis the measurement and characterization of what constitutes a social link remains still an unanswered issue. As mentioned above, this is mainly due to the fact that networks of electronic communication significantly differ from the physical one in one substantial thing: by quoting Tang et al. (2011) "physical social networks are colorful ('family members', 'colleagues', and 'classmates')" while when looking at electronic networks they are usually "black-and-white", in the sense that no information is given by the activity data on the type of the underlying social tie. To give a representation of social ties close to the real one, one usually keeps only ties that are reciprocated and assign them a weight that should reflect its importance and nature in real life (see Sect. 2.1.2). Of course there are several ways to assess tie reciprocity and as much in which tie weights can be defined and assigned. In line with aggregated and static approaches traditionally used to model social networks, one usually assesses the reciprocal character of a tie between two individuals depending on whether there has been at least one reciprocated pair of communication events between them during the whole observation time window under investigation. Within the same picture, communication tie weights are usually taken as the total volume or intensity of communication between any pair of individuals in the whole time period (e.g. number or duration of calls in mobile-phone networks or number of directed messages in Twitter) (Huberman et al. 2009; Onnela et al. 2007b). Indeed, it has been shown that the volume of communication correlates with the importance of the face-to-face relationship (Eagle et al. 2009). As we will see in this section, the interpretation of tie strength leads to observation and validation of many structural properties that were already known from smaller and/or face-to-face networks (Baym et al. 2004; Wellman and Haythornthwaite 2003). As a particular case of social networks, communication networks present many of the topological properties described in Sect. 2.1: large heterogeneity in both the social connectivity and the strength of nodes and ties; non-trivial clustering or network transitivity; positive correlations or assortative mixing between the degrees of adjacent vertices; they are often divided into groups or communities that, as it has recently been suggested by Newman (2003a), may account for the observed clustering. The aim of this section is to present some of the most outstanding results about communication networks known in literature, with a particular focus on all those topological aspects which make them different from other social networks. All the results presented have been obtained by considering a static snapshot of the network, resulting from the aggregation of all communication events over a given time window. There are however several limitations to this description, starting from the definition of tie weight as the mere volume of communication, limitations that will be discussed in more details in Sect. 2.3.

2.2.1 Topological Properties

As discussed in Sect. 2.1, a basic network characteristic is the degree distribution. We have seen that many social networks show in general a skewed degree distribution which follows a power-law behavior $P(k) \sim k^{-\gamma}$ with exponent γ between 2 and 3, indicating the existence of hubs or people with a very large number of connections. In this aspect, communication networks slightly differ from other social networks. For e-mail communication networks, for example, it has been observed that the distribution of the number k of a node's next neighbors obeys an exponential behavior $P(k) \propto \exp(-k)$ (Guimerá et al. 2003). According to other studies (Amaral et al. 2000; Newman et al. 2002), the truncation of the scale-free behavior in real world networks is due to the physical costs of adding ties and the limited capacity of an individual (Bonney 1956). Other results indicate instead a heavily skewed degree distributions for e-mail where a power-law $P(k) \sim k^{\gamma}$ with an exponent $\gamma \sim -1.8$ (Ebel et al. 2002; Ferrara 2012). Small deviations from a power-law behavior with $\gamma \in [2, 3]$ have been observed also for online social networks, where in general two different regimes are observed for $P(k)$: a rapid decay ($\gamma \sim 4 - 5$) for small k and a heavy tailed ($\gamma \sim 1 - 2$) for large values of k (Ahn et al. 2007; Ferrara 2012; Kwak et al. 2010; Mislove et al. 2007). The range of large k for this type of networks is usually associated to atypical users, i.e. those individuals with a large audience whose messages get broadcasted. A rapid decay of $P(k)$ has also been observed in mobile phone communication networks where, although the tail of the degree distribution is better approximated by a power-law that an exponential, the obtained exponent is significantly higher. Onnela et al. (2007a) for example, found an exponent $\gamma = 8.4$, in which cases the power-law distribution can be easily confused with exponential (Clauset et al. 2009). Despite the differences, most results for communication networks show however a decay in the degree distribution which is faster than a power-law, indicating that the hubs are few. In phone communication networks this decay is probably due to the fact that business numbers are usually filtered out from the analysis and that each event usually represents a one-to-one communication (Miritello et al. 2012b; Onnela et al. 2007b), in contrast with e-mail or instant messaging networks, in which the recipients of the message can be many and well-connected hubs are observed (Ebel et al. 2002). Other times the observed differences may depend on the way in which interaction ties are defined. In networks as Twitter, for example, the social graph resulting from the following/follower relations can significantly change from the interaction one, where a tie between two persons is considered only if there has been direct communication between them (Huberman et al. 2009).

Consistently with general results about social networks, also communication networks show a well organized structure typically characterized by the existence of communities of individuals. Since the existence of communities between individuals is intuitively clear in human society, it is not surprising to find communities also in all those networks that represent interactions between individuals. Generally, the distribution of community sizes shows a slow decay, indicating that there is no characteristic group size (Ferrara 2011; Grabowicz et al. 2012; Guimerá et al. 2003). In large online social networks such as Facebook however, it has been found that there is a high probability of finding a high number of communities that contain few individuals and a lower probability of finding communities constituted by a large number of members (Ferrara 2012; Leskovec et al. 2009). The latter result suggests that individuals are more likely to aggregate in small communities, such as those representing family, friends or colleagues, rather than in large communities. On the other hand, for some networks such as e-mail or mobile phone networks, results show that there exists an important number of communities with a large amount of individuals, constituting the heavy long tail of the observed power law distribution (Blondel et al. 2008). As well as for other social networks, also in communication networks the existence of a community structure can be associated to the assortative mixing of some attribute of the vertices. As mentioned in Sect. 2.1, social networks are assortative: people with many friends are connected to others who also have many friends. This has been observed for online communication networks (Kwak et al. 2010) as well as for mobile-phone networks (Onnela et al. 2007a). This human tendency to interact with individuals similar to themselves is also know as *affinity* or *homophily* and emerges not only in the degree, but also with respect to a variety of attributes from psychological states such as loneliness or happiness (Bliss et al. 2012; McPherson et al. 2001) or health attributes and habits (Christakis and Fowler 2007, 2008), to tastes and interests (Lazarsfeld and Merton 1954; Lewis et al. 2008) and sociodemographic features such as age or race (Ibarra 2002; Mollica et al. 2003). As we will show in Chap. 3, the inherent homophilous nature of humans not only emerges in topological or psychological and exogenous factors, but also in dynamical processes of human interaction which have not been thoroughly investigated so far (Miritello et al. 2012a).

Since the nodes of a network may have positions in space, in many cases, it is reasonable to assume that geographical proximity plays a role in deciding how to connect the nodes, something that has indeed been observed for many real networks (Barthélemy et al. 2003; Boguña et al. 2004; Kleinberg 2000). For example, in mobile phone and online social networks it has been observed that the probability for two individuals to be connected decays with their geographical distance (Lambiotte et al. 2008; Onnela et al. 2011). On the other hand, most of these networks also appear to be geographically disperse (Barthélemy 2003; Lambiotte et al. 2008; Kwak et al. 2010). This result can be easily understood since, although the geographical proximity is an essential condition for face-to-face interactions, this is not true for the majority of communication networks whose function is indeed to enable the interaction with people who do not necessarily share the same physical place. Despite their geographical dispersion, however, communication networks are highly connected in terms of graph-distance and they appear to be even smaller than other social networks. A very recent study of Facebook interaction network show for example that the average number of intermediate ties between two randomly chosen humans (shortest path length) is almost 4 (Backstrom et al. 2012). This value, which is significantly smaller than the 6-degrees found in the original experiment by Milgram (Milgram 1967), indicates that, when considering another person in the world, "a friend of your friend knows a friend of their friend, on average" (Backstrom et al. 2012).

2.2.2 *Correlation Between Topological Structure and Tie Weights*

Topological Overlap and the Strength of Weak Ties

As mentioned in the previous section, people tend to form groups with other people similar to themselves. This suggests that in general ties within communities have different properties than ties connecting the communities (bridges). One of the most known result in this respect, hypothesized by the american sociologist Mark Granovetter (Granovetter 1973), is that ties within communities tend to be stronger that the ones between them (Onnela et al. 2007b; Lewis et al. 2008). This hypothesis is known as the *weak ties hypothesis* and implies the existence of important correlations between local network structure at the level of communities, and interactions strengths. One way to measure this correlation is by looking at the relation between the tie strength w_{ij} between two nodes i and j in a network and the relative topological overlap of their common neighbors, defined as:

$$O_{ij} = \frac{n_{ij}}{(k_i - 1) + (k_j - 1) - n_{ij}}, \tag{2.11}$$

where k_i and k_j are respectively the degrees of the two nodes and n_{ij} the number of neighbors common to both of them (Onnela et al. 2007b). If i and j have no common acquaintances, then $O_{ij} = 0$ and the tie between the two nodes represents a potential bridge between two different communities. On the contrary, if i and j are part of exactly the same circle of friends, then $O_{ij} = 1$. A positive correlation between w_{ij} and O_{ij} has been observed, for example, in mobile communication networks (Onnela et al. 2007b). By defining w_{ij} as the total volume of communication between i and j over a given time window, Onnela et al. found that the more the time two individuals spend talking together is, the more their friends overlap or the other way around, which is in line with the strength of weak ties hypothesis. As a consequence, the network structure in the vicinity of a randomly selected individual is similar to the one depicted in Fig. 2.3: consistent to the weak ties hypothesis, the majority of strong ties are found within the clusters, while most links connecting different communities are much weaker. An alternative measure of the topological overlap is given by the Adamic-Adar index s^{AA} which refines the simple counting of common neighbors by giving the lower-connected neighbors more weights (Adamic and Adar 2003). It is defined as $s_{ij}^{AA} = \sum_{z \in N(i) \cap N(j)} 1/\log(k(z))$, where $N(i)$ and $N(j)$ are, respectively, the neighbors of i and j.

The structural configuration of a network can have global implications on its stability and functionality since links may have a different role or function in the system depending on their strength and/or their location with respect to the groups.

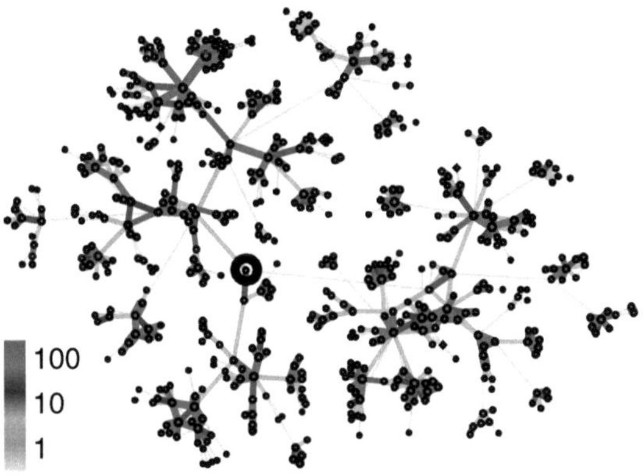

Fig. 2.3 Structure of a mobile phone network around a randomly selected individual (marked by the *black circle*), where only nodes which are at distance less that six from the selected one are shown. Each tie represents a reciprocated tie (mutual calls) between the involved users and tie weight is defined as the aggregated call duration in minutes (see *color bar*). Adapted from "Structure and tie strengths in mobile communication networks", Onnela et al. (2007b)

It has been observed, for example, that weaker ties are crucial for maintaining the network's structural integrity and that a removal of few of them from the whole network drives the system into a rapid disintegration. On the contrary, given that strong ties are predominantly within the communities, their removal only disintegrate a community but does not affect the overall integrity of the network (Onnela et al. 2007b).

This finding shows a significant difference between social networks and biological or technological ones, where exactly the opposite is observed and immediate network's collapse is caused by the removal of strong links (Barthélemy et al. 2004). Weak ties play an important role also in the dissemination of information within a network, since they help to link together different parts of a system while strong ties significantly slow information flow, trapping it in communities (Onnela et al. 2007b). Recently it has been proven that the weak ties hypothesis also applies to online networks (Grabowicz et al. 2012; Ferrara 2011). For example, by defining the strength of a tie ij as the total number of personal messages exchanged between i and j over the period of observation, Grabowicz et al. (2012) found that weak links are typically connections between persons not sharing neighbors and also in this case they contribute to more efficient information flow. In fact, while personal messages tend to concentrate inside the communities, retweets, which are associated to information propagation events, appear with higher probability in links between groups.

Dunbar's Number

Despite the large size of many real-world communication networks and their small diameter, they appear to organize in relatively small size communities of 50–200 individuals (Leskovec et al. 2009; Ferrara 2011). This observation agrees well with the so-called *Dunbar's theory* which predicts that the cognitive limit to the number of people with whom each person can have a close relationship is roughly 150. This number takes its name from the anthropologist Robin Dunbar who in 1992 measured the correlation between neocortical volume and typically social group size in a wide range of primates and human communities (Dunbar 1992). In real-world social interactions, beside the biological constraints, also other physical limitation may play their role, first of all the fact that time and attention are scarce resources and people have a finite amount of it to dedicate to social relations. Recently, Dunbar's theory has been tested with regard to Twitter users, and it was found that the new mode of communication did not have a significant impact on the human biological and cognitive limits to social interactions, with Twitter users maintaining a maximum of 100–200 stable relationships (Gonçalves et al. 2011). The Dunbar's theory also asserts that numbers of social relationships larger than 100–200 have generally a higher cost and require more effort to maintain a stable connection. This is in line with that found in Grabowicz et al. (2012) with regard to Twitter users: despite the existence of communities with size significantly larger than Dunbar's number, links with direct messages are much more abundant within groups of size up to 150 users. A similar result has been observed in Facebook communication where, despite the large number of declared friends, users only poke and message a small number of people (Golder et al. 2007).

As we will show in Chap. 3, we obtain similar results for mobile phone communication networks. We observe however a slightly smaller limit of the social capacity, probably because beside the cognitive limit, also temporal and monetary constraints play their role in phone communication (Miritello et al. 2012b).

2.3 Traditional Network Modeling

Up to now we have seen how to characterize the structure of a network, focusing on how the main properties of its elements have been traditionally defined and measured. We have also seen that in some cases the simple modeling framework can be made more powerful by including additional levels of details, as for example ties weights, whose correlations with the network topology give also important insights on how people behave, interact and organize. In the last years, the understanding of the structure of many communication networks has received huge interest among the scientific community and a lot of literature has been produced based on the analysis of phone call networks (Akoglu and Dalvi 2010; Hidalgo and Rodriguez-Sickert 2008; Nanavati et al. 2008; Onnela et al. 2007b), e-mail networks (Ebel et al. 2002; Eckmann et al. 2004; Guimerá et al. 2006; Kossinets and Watts 2006) and online social networks (Ahn et al. 2007; Ferrara 2012; Mislove et al. 2007; Ugander et al. 2011; Kwak et al. 2010; Huberman et al. 2009). The majority of studies on social networks structure have focused mainly on *(i)* the way real networks deviate from completely random networks, *(ii)* how the observed structure can emerge from individual behavior and *(iii)* how the observed network topology can be modeled. This type of analysis constitutes the first step to characterize network topology, understand which nodes play a similar role in a given system and the way in which they are connected to each other. Characterizing network topology not only helps in the understanding of how a part of the network differs from others, but also in the description and modeling of dynamical processes such as information spreading (Iribarren and Moro 2011a; Ugander et al. 2011) or influence (Aral and Walker 2012). All these processes, in fact, are constrained by the way in which whom and how each individual is connected to and located within the network. The fact that in social networks the flux of information that passes through each tie is unevenly distributed, some individuals are much more connected than others and social relationships are organized in communities, must indeed reflect in the way in which information, opinions or influence spread.

Paradoxically, the majority of these studies are based on aggregated statistics and ignore one inherent property of real social networks: the fact that they tend to change dynamically. Most social and communication networks are in fact *temporal networks*, not only in the sense that they are subject to a continuous evolution over time, but also because interactions between agents happen at a given time and may have a given duration (Holme and Saramäki 2012). Traditional network models are instead essentially static and all information about the time at which social interactions take place is discarded. The contact network is in fact obtained by aggregating over time all the interaction events observed in a given time window. This representation results therefore in a static snapshot of social interactions where the temporal dimension is completely projected out. In this representation all nodes and ties are considered as appearing at the same time and are assumed continuously active. In addition, the nature of social relationships reduces to a static strength which, although it incorporates the volume of communication between two individuals, it does not include any information about the way in which the involved individuals interact in time. The implications of these assumptions are very strong since they imply that people can interact with the same probability with any other individual in their social circle and that interaction can happen at any time and that there is no causal correlation between interaction events.

In contrast, everyday life experience suggests that none of these assumptions actually holds: people do not communicate everyday with any other people they know; interaction with one of our friends or colleagues may trigger the short-term interaction with other people; while some social relationship can last for years, others are very short in time or more occasional. Therefore, although the volume of communication may be an indicator of the importance or the role that a particular person has in our life, it does not capture any information on the temporal duration of a social tie or the way in which such a volume of interaction events is distributed within a given time window. In Chap. 4 we will show that, actually, the same tie strength can correspond to ties with very different temporal features and that static approaches thus radically misrepresent the real patterns of interactions and miss important aspects and tendencies of dynamical networks. As we will see in Chap. 3 and 5, the ongoing change of the networks may affect the instantaneous structure of the network, playing a fundamental role in the evolution of communities

(Palla et al. 2007a; Tantipathananand et al. 2007) and individual's communication strategies (Miritello et al. 2012a) and the way in which communication events are distributed in time is crucial in spreading phenomena (Iribarren and Moro 2009; Karsai et al. 2011; Miritello et al. 2011; Rocha et al. 2010; Vázquez et al. 2007).

One of the standard approaches to take into account temporal properties of interaction when modeling temporal networks has been to divide the time period under consideration into smaller time windows, then aggregate the social network in each window separately (Sarkar and Moore 2005; Snijders 2001). These approaches are, however, still static since they do not account for the fact that social relationships are mostly instantiated intermittently over time. An alternative method is given by the *time-respecting* graph where a tie (i, j) is defined as a time-labelled tie (i, j, t), where paths need to obey the time order of the appearance of ties (Kempe et al. 2002). According to this model however, temporally disconnected nodes are not considered and the frequency of contacts between nodes is not taken into account. A similar approach has been proposed where nodes, instead of ties, are labeled at each time instant they appear and, whenever a connection between two nodes is observed, a link between them is established, with weight equal to the time difference between the nodes' time appearances (Kostakos 2009). The main problem of these approaches is however the fact that, as we will show in Chap. 3, due to the strong heterogeneities of human interactions, nodes and/or ties activity can be confused with their appearance and, although no activity between i and j is observed at time t, a connection between them can, instead, exist. Other approaches use the concept of *reachability* to define temporal distance metrics where a directed tie from i to j is considered if there is a time-respecting path from i to j (Moody 2002; Tang et al. 2009, 2010). In contrast to others, the latter methods are able to capture the duration and time order of contact and have been show to be useful to quantify information diffusion processes (Holme 2005; Tang et al. 2009). To take into account the time between two consecutive contact on a path, generalizations of time-respecting path approaches which set a limit to the maximum allowed waiting time at a node have been also proposed (Pan and Saramäki 2011). All these methods certainly represent an improvement over the static aggregated approaches. However, a general frame of how to model dynamical social networks by taking into account both topological and temporal aspects of human behavior, as well as the correlations between them, is still lacking. The main reason why the temporal dimension has been neglected in traditional network modeling is certainly due to the fact that it is usually much easier to analyze static graphs. However, there are at least two other reasons why the underlying static network and the dynamical system usually appear separated and little is known about how to model temporal networks and their dynamical processes. The first reason, addressed in Chap. 3, is based on the belief that temporal processes happen very slowly such that the structure of the network is not remarkably affected. Although in many cases the temporal dimension can indeed be too insignificant over the periods of study used to be included in the network analysis, it can be of paramount importance in other cases. The second reason is associated to the lack of longitudinal data until few years ago. In fact, as mentioned above, the interest in modeling dynamically temporal networks has enormously increased only in the last few years with the rapid appearance of fine grained electronic longitudinal data, such as phone-communication records, emails, web, blogs and online social networks. The availability of this data has sparked numerous investigations into not only the topological, but also into the temporal properties of human interactions (Barabási 2005; Gaito et al. 2012; Goh and Barabási 2008; Kleinberg 2008; Kossinets et al. 2008; Kovanen et al. 2011; Jo et al. 2012; Rybski et al. 2010). What emerges is that temporal patterns of human interaction are actually very complex and articulated to be neglected in the description and characterization of social networks. Actually, as we will see in the next section, the inherent properties of temporal activity patterns also play a crucial role in all those processes where the temporal ordering of events is important, such as spreading phenomena, emergence of collective behavior, opinion formation or human synchronization, indicating that traditional models of social networks need to be revised.

2.4 Temporal Properties of Social Networks

As we have seen in the previous section, static descriptions of networks usually neglect the temporal dimension of human activity. Among others, some of the implicit assumption of this approach are that *(i)* nodes, edges, communities do not change in time, *(ii)* human actions are markovian and randomly distributed in time, therefore well approximated by Poisson processes, and *(iii)* there is no correlation or causality between interaction events. In recent years, however, there has been an increasing evidence that none of these hypothesis applies for real social networks. In this section we will see that these recent findings show that the dynamics of social networks is much more articulated and evolves as, contrary to static descriptions, social interactions are dynamical, tie decay/form and nodes enter/exit the social network. Since the understanding of how each of these properties affect the current way of modeling social networks is the main goal of this thesis, we will also discuss some of the implications that temporal patterns of communication have on the static description of the underlying contact network. This should serve to the reader as a preliminary baseline for all the results presented in rest of this work.

2.4.1 Nodes and Ties are Not Persistent

One of the basic assumption of traditional networks models is that nodes and ties within a network are continuously active. However, in the majority of real systems they are not since people may join and leave the network over time (Hidalgo and Rodriguez-Sickert 2008; Kossinets and Watts 2006). This makes the number of total nodes in the network not constant in time, thus changing the whole network structure and its properties (Ebel et al. 2002). For example, in co-authorship networks new authors can appear or abandon the network over time (Barabási et al. 2002), an effect

that has also be observed in other networks as internet dating (Holme et al. 2004) and mobile communication networks (Palla et al. 2007a). As well as individuals, also social relations are not always active in time and are characterized by very different lifetimes. As we will see in Chap. 4, while some relationship is observed only for the duration of an interaction, others may last beyond the interaction with which they began. In some cases this is due to the very definition of a link: since interactions usually have a given duration in time, ties activation and deactivation reflects the continuous changes in the activity and communication patterns of individuals. In all these cases, the instantaneous picture of the network can be significantly different from the aggregated one, where all the nodes and interactions observed in the entire observation period are taken into account. Moreover, the fact that individuals activate and deactivate social ties over time not only alters the structure of the networks in which they participate but also affect the dynamics of many phenomena that happen through the network, from communities formation (Palla et al. 2007a; Tantipathananand et al. 2007) to spreading processes (Iribarren and Moro 2009; Karsai et al. 2011; Miritello et al. 2011). There are many forces that govern the activation/deactivation of social relationships and the tendency for relations to weaken and disappear. The problem to identify under what conditions some ties are more likely to dissolve or persist, which will be the main focus of Chap. 4, is know in the literature as the *link prediction* problem and it has been the objective of many studies in recent years. The most exemplary work on edge decay in social networks is probably that of Burt who studied the social networks of the most important bankers over time and analyzed those factors that contribute to the disappearance of edges between them (Burt 2000, 2002) . Several other studies have focused on link characterization in other communication networks as phone and SMS networks (Akoglu and Dalvi 2010; Hidalgo and Rodriguez-Sickert 2008; Raeder et al. 2011), online networks (Aiello et al. 2010; Crandall et al. 2008; Gilbert and Karahalios 2009; Kivran-Swaine et al. 2011; Romero et al. 2011) and off-line settings (Burt 2000; Martin and Yeung 2006).
One of the main substantive conclusions that emerge from these studies is that links exhibit a memory, meaning that old links are more likely to persist in time than newly-formed ones (where the "age" and persistence of a link is defined in terms of observed communication events or when the link is registered or deleted in on-line social networks). Several other factors influence edge decay/persistence including *structural* properties of ties as reciprocation, neighborhood overlap or clustering coefficient, *temporal* properties as the time since the last communication (Raeder et al. 2011), homophily (Crandall et al. 2008) or *geographical* aspects as individual location and distance (Liben-Nowell et al. 2005; Lee et al. 2009). Tie creation or decay may also signal changes in the community structure and the condition for stability for large communities is strictly related to the continuous changes in their membership (Palla et al. 2007a).

Taking into account the ongoing appearance and disappearance of nodes and edges in the modeling of real social networks can lead to a characterization that significantly differs from the aggregated static one. In Chap. 3, for example, we will show that, since many social interactions are not always active over time, the instantaneous number of connections of an individual is actually much smaller than the time-aggregated one. This indicates that the standard social connectivity (defined in Sect. 2.1.1) usually overestimates the actual peoples social capacity of maintaining ties. We will show that according to the volatility of social relationships, it is possible to identify different individual communication strategies, from exploratory to stable. Depending on the adopted strategy, individuals also play a different role in spreading phenomena. Again, the dynamics of tie creation/removal is crucial since, in contrast, aggregated models assume that all connections are equally stable over time and that all users have the same communication strategy. Due to the importance that both social connectivity and its correlations with tie weights have on the characterization of network topology and on the modeling of many real phenomena as information spreading or network resilience, the consequence of accounting or not for the ongoing dynamics of human interactions may be therefore considerable.

2.4.2 Inter-Event Times and Bursty Behavior

As mentioned above, one of the implications of considering a static and aggregated snapshot of temporal interactions is that human interactions are randomly distributed in time, thus well approximated by homogeneous and memoryless Poisson processes (Greene 1997; Haight 1967; Reynolds 2003). Homogeneous Poisson processes have two main statistical properties: the number of events during a time interval of duration T follows a Poisson distribution with mean ρT and the time δt between consecutive events, called the *inter-event* or *waiting* times, follows an exponential distribution $p(\delta t) \sim \rho \exp(-\rho \delta t)$. As a consequence, individual actions happen at relatively regular time intervals δt and very short or very long inter-event times occur with small probability (see Fig. 2.4a). However, in real networks, this is usually not the case.

One of the main results that has emerged in the last few years from observing social interactions is that temporal patterns of human individuals are strongly inhomogeneous. This is reflected in the slowly decaying of the inter-event time distribution, which has been found to have a heavy tail with a power-law decay as $P(\delta t) \sim \delta t^{-\gamma}$, with $\gamma \simeq 1$ (Barabási 2005). As shown schematically in Fig. 2.4, the latter is in stark contrast with the prediction of a homogeneous Poisson process. This behavior seems to be a universal feature of human activity. It has in fact been observed in several systems driven by human activity sequences (Barabási 2010; Eckmann et al. 2004; Goh and Barabási 2008; Oliveira and Barabási 2005; Rybski et al. 2010) and is known in literature as *bursty behavior* since long periods of inactivity are separated by intense bursts of activity. According to other studies (Karsai et al. 2011; Rybski et al. 2010), in Chap. 5 we will show that bursty behavior is observed not only in the way an individual executes tasks, but also in the interaction between two individuals (Miritello et al. 2011). The heavy-tailed nature of the distribution of inter-event times generated a huge interest in the last years in the understanding what are the mechanisms responsible for its emergence. Two main classes of mechanisms have

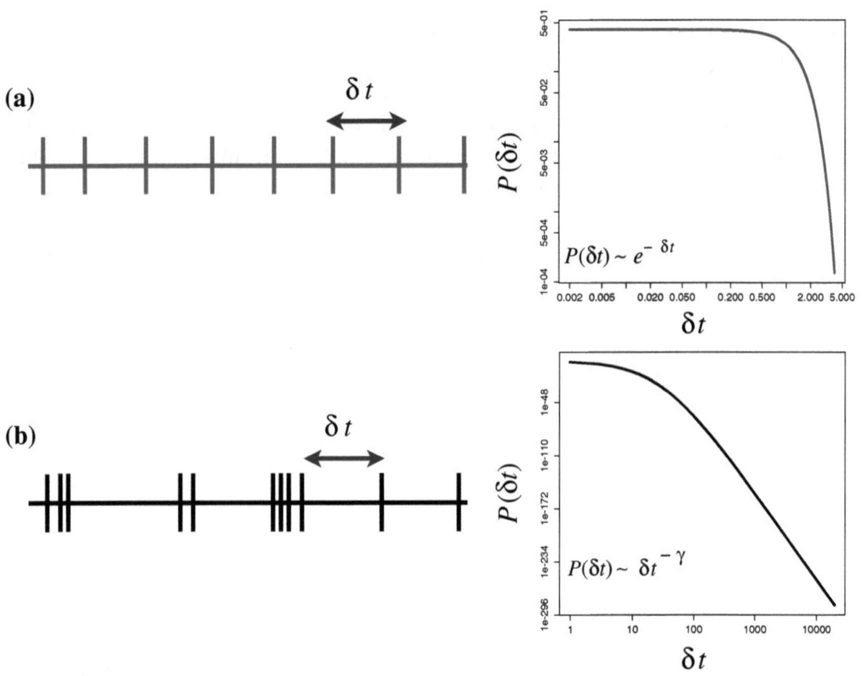

Fig. 2.4 Schematic representation of the difference between the homogeneous activity pattern predicted by a Poisson process and the heterogeneous behavior observed in human dynamics. **a** According to a Poissonian process the events are homogeneously distributed in time, thus the inter-event time δt follows an exponential distribution. **b** Bursty pattern of activity observed in real systems, in which bursts of events are followed by periods of inactivity giving rise to an heavy-tailed distribution for the inter-event time δt

been suggested: *(i)* human behavior is driven by external factors such as circadian and weekly cycles, which introduce a set of different time scales that give rise to the heavy tails (Malmgren et al. 2008); *(ii)* temporal inhomogeneities are caused from human task execution behavior, which is driven by a priority selection mechanism which introduce correlations in activity (Barabási 2005; Vázquez et al. 2006; Waleaevens et al. 2012). This latter mechanism is supported by a recent finding by Jo et al. (2012) who, after removing circadian and weekly patterns in the time-series by applying de-seasoning methods, observe the robustness of the inter-event times distribution of mobile phone communication events of individuals. The bursty dynamics of tie interactions has very important and at times drastic effects in the characterization of the underlying social network. For example, one of the implications of a Poissonian description of human dynamics is that social ties are characterized only by the number of communication events between the involved individuals, which regulates the rate of the process. This implies that all ties having the same weight are equivalent. In contrast, as we will show in Chap. 4, ties with exactly the same number of communication events can be instead characterized by a very different distribution of these events within their lifetime period. The bursty dynamics of human interactions, together with the temporal correlations between interactions, also have significant implications on many dynamical phenomena happening on the network in which the time ordering and delay is crucial, such as spreading processes. In recent experiments of electronic recommendation, indeed, it has been shown that the large heterogeneity found in the waiting times is responsible for the slow dynamics of information at the collective level (Iribarren and Moro 2009), which makes the observed dynamics significantly different from the Poissonian expectations (Iribarren and Moro 2011b; Karsai et al. 2011; Miritello et al. 2011; Vázquez et al. 2007). The latter topic will be discussed in details in Chap. 5.

2.4.3 Temporal Correlations: Motifs and Group Conversations

In Sect. 2.1.1 we have seen that the topology of many social networks is characterized by motifs, e.g. patterns of interconnections that appear with significantly high frequency. When dealing with temporal networks, however, these patterns can be affected by the temporal order at which interactions take place. Several ways have been proposed to extend the concept of motifs in order to take into account the changes in the network structure over time. Some of these consider snapshots of the network at different times, then look at the aggregated ties in each sub-window and count the different networks in these snapshots (Braha and Bar-Yam 2008), while others use the temporal information for defining the subgraphs of interest (Kossinets and Watts 2006; Kovanen et al. 2011; Zhao and Oliver 2010). The latter approach gives a more precise description of real temporal motifs since the sequences of events are based on temporal order of the events, instead of considering aggregated snapshot windows. The results show that the number of such paths is significantly larger when compared with reference networks, where the event times have been randomly reshuffled. Understanding recurrent temporally ordered patterns might yield a lot of insights on network analysis, especially in social and interaction networks, since it reveals the existence of correlations between patterns of communication, which may have a causal explanation. In communication networks, the presence of correlations between events has also been observed by looking at the distribution of the *relay time* τ_{ij}, also called *inter-contact time*, which measures the time it takes for the individual i to interact with j after having interacted with any other person $* \neq j$ (Cattuto et al. 2010; Eckmann et al. 2004; Isella et al. 2011; Miritello et al. 2011; Wu et al. 2010; Zhao and Oliver 2010). The relay time τ_{ij} depends not only on the inter-event times δt_{ij} in the $i \leftrightarrow j$ communication, but also on the possible correlations with the $* \leftrightarrow i$ events (Newman 2002b), thus on the way in which group conversations happen. In a first approximation, where the latter correlation is neglected, the relation between τ_{ij} and δt_{ij} is given by the waiting time density for δt_{ij}:

$$P(\tau_{ij}) = \frac{1}{\overline{\delta t}_{ij}} \int_{\tau_{ij}}^{\infty} P(\delta t_{ij})\, d\delta t_{ij}, \tag{2.12}$$

where $\overline{\delta t}_{ij}$ is the average inter-event time. The heavy-tail properties of the distribution $P(\delta t_{ij})$ of inter-event times are therefore inherited by $P(\tau_{ij})$, which appears to be skewed with a long-tail (Miritello et al. 2011; Karsai et al. 2011; Rybski et al. 2009). Interestingly, the results for $P(\tau)$ also show that not only large, but also very small relay times are much more probable when compared with the series of time-reshuffled events. This reveals indeed the existence of group conversations between individuals. As well as the other temporal inhomogeneities of human interactions described above, also correlations between events have crucial implications on the dynamics of real processes. Indeed, in Chap. 5 we will show that, together with the bursty behavior, group conversations are the main dynamical ingredient in the understanding of the spread of information in social networks and the principal responsible for the observed disagreement between Poissonian expectations and empirical results (Miritello et al. 2011).

2.5 Discussion

The purpose of this chapter has been to introduce the basic concepts and the main properties of social networks, with a particular interest on communication networks, which are the focus of this thesis. In communication networks each node usually represents an individual and the ties between individuals represent one or more communication events between them, such as phone-call, e-mail or online message communication.

We have seen that these networks, as well as many other social networks, are characterized by a very heterogeneous structure: some individual is much more connected than others, the flux of information which passes through social connection is not evenly distributed and social relationships are organized within communities. We have also presented how all these properties can be measured and analyzed in order to get insights on how people act and interact and have discussed the traditional way in which social networks have been modeled. We have seen that most of the analysis present in the literature has been restricted on characterizing the network topological structure in a given observation time window and on modeling its dynamics. In this analysis, real networks have been usually modeled by following what we call static approaches, where the temporal dimension is completely projected out: nodes and relationships between them are considered active during the whole period under investigation, interactions are basically characterized by the volume of interaction between the two end-nodes and can happen at any time and that the communication patterns are homogeneous in time, thus well approximated by Poissonian processes.

In contrast, many real systems are temporal networks, characterized by non trivial temporal patterns (Holme and Saramäki 2012). We have seen, in fact, that in many social networks relationships appear and disappear in time, leading to an instantaneous contact network which may be different from the one that emerges when they are considered everlasting (Hidalgo and Rodriguez-Sickert 2008; Kossinets and Watts 2006; Palla et al. 2007b). We have also seen that interaction events are correlated and human actions are bursty (Barabási 2005; Eckmann et al. 2004; Kossinets and Watts 2006; Vázquez et al. 2006) thus can not be modeled by homogeneous Poissonian processes. All these properties, which can not be captured by static network models, are emerging only in the very last years, thanks to the increasing availability of large electronic data sets of human communication that contain fine-grained temporal information of how people act and interact. We have also mentioned that, as well as topological features of human interaction, temporal properties are crucial to understand and characterize the underlying social network and the dynamical processes happening through it (Iribarren and Moro 2009; Vázquez et al. 2007), something that we will analyze in more details in the next chapters.

References

Adamic L, Adar E (2003) Friends and neighbors on the web. Soc Netw 25(3):211–230

Ahn Y-Y, Han SK, Moon S, Jeong H (2007) Analysis of topological characteristics of huge online social networking services. In: Proceedings of the 16th international conference on the, World Wide Web, pp 835–844.

Ahuja RK, Magnanti TL, Orlin JB (1993) Network flows: theory, algorithms, and applications. Upper Saddle River, New Jersey

Aiello L, Barrat A, Cattuto C, Ruffo G, Schifanella R (2010) Link creation and profile alignment in the aNobii social network. In: Proceedings of the sec- ond IEEE international conference on social computing socialCom 2010, Minneapolis, USA.

Akoglu L, Dalvi B (2010) Structure, tie persistence and event detection in large phone and SMS networks. In: MLG '10 Proceedings of the eighth workshop on mining and learning with graphs.

Albert R, Jeong H, Barabási A-L (1999) The diameter of the world wide web. Nature 491:130–131

Albert R, Barabási A-L (2002) Statistical mechanics of complex networks. Rev Mod Phys 74:47–97

Almaas E, Kovacs B, Viscek T, Oltvai Z, Barabási A-L (2004) Global organization of metabolic fluxes in the bacterium Escherichia coli. Nature 427:839

Amaral LAN, Scala A, Barthélemy M, Stanley H (2000) Classes of small-world networks. Proc Natl Acad Sci U S A 97:11149–11152

Anderson RM, May R (1992) Infectious diseases in humans. Oxford University Press, Oxford

Aral S, Walker D (2012) Identifying influential and susceptible members of social networks. Science 337:337–341

Aral S, Van Alsyne M (2007) Network structure and information advantage. Proceeding of the academy of management conference, Philadelphia, PA, In

Arenas A, Duch J, Gómez S, Danon L, Díaz-Guilera A (2010) Communities in complex networks: Identification at different levels. In: Caldarelli G (ed) Enciclopedia of life support system (EOLSS). EOLSS Publishers, Oxford, UK, Developed under the auspices of the Unesco

Backstrom L, Boldi P, Rosa M, Ugander J, Vigna S (2012) Four degrees of separation. arXiv:1111.4570v3.

Barabási A-L (2005) The origin of bursts and heavy tails in human dynamics. Nat Sci Rep 435:207–211

Barabási A-L (2007) The architecture of complexity. IEEE Control Syst Mag 27:33–42

Barabási A-L (2010) Bursts: the hidden pattern behind everything we do. Dutton Books, New York

Barabási A-L, Albert R (1999) Emergence of scaling in random networks. Science 286:509–512
Barabási A-L, Jeong H, Ravasz R, Neda Z, Vicsek T, Schubert A (2002) Evolution of the social network of scientific collaborations. Physica A 311:590–614
Barrat A, Barthélemy M, Pastor-Satorras R, Vespignani A (2004) The architecture of complex weighted networks. Proc Natl Acad Sci U S A 101:3747
Barthélemy M (2003) Crossover from scale-free to spatial networks. Europhys Lett 6:915
Barthélemy M, Gondran B, Guichard E (2003) Spatial structure of the internet traffic. Physica A 319:633642
Barthélemy M, Barrat A, Pastor-Satorras R, Vespignani A (2004) Velocity and hierarchical spread of epidemic outbreaks in scale-free networks. Phys Rev Lett 92:178701
Barthélemy M, Barrat A, Pastor-Satorras R, Vespignani A (2005) Characterization and modelling of weighted networks. Physica A 346:34–43
Baym NK, Zhang YB, Lin M (2004) Social interactions across media: interpersonal communication on the internet, face-to-face, and the telephone. New Media Soc 6:299
Bliss C, Kloumann I, Harris KD, Danforth C (2012) Twitter reciprocal reply networks exhibit assortativity with respect to happiness. J Comput Sci 3:388–397
Blondel V, Guillaume J-L, Lambiotte R, Lefebvre E (2008) Fast unfolding of communities in large networks. J Stat Mech, P10008
Boccaletti S, Latora V, Moreno Y, Chavez M, Hwang D-U (2006) Complex networks: structure and dynamics. Phys Rep 424:175
Boguña M, Pastor-Satorras R, Díaz-Guilera A, Arenas A (2004) Models of social networks based on social distance attachment. Phys Rev E 70:056122
Bollobás B (1985) Random graphs. Academic, New York
Bollobás B, Riordan O, Spencer J, Tusnady G (2001) The degree sequence of a scale-free random graph process. Random Struct Alg 18:279–290
Bonney M (1956) Sociometry and the science of man. Beacon House, New York
Braha D, Bar-Yam Y (2008) Time-dependent complex networks: dynamic centrality, dynamic motifs, and cycles of social interaction. In: Gross T, Sayama H (eds) Adaptative networks: Theory, models and applications. Springer, Dordrecht, pp 39–50
Burt R (2000) Decay functions. Soc Netw 22:1–28
Burt R (2002) Bridge decay. Soc Netw 24:333–363
Callaway D, Newman MEJ, Strogatz SH, Watts DJ (2000) Network robustness and fragility: percolation on random graphs. Phys Rev E 85:5468–5471
Candia J, González M, Wang P, Schoenharl T, Madey G, Barabási A-L (2008) Uncovering individual and collective human dynamics from mobile phone records. J Phys A: Math Theor 41:224015
Cattuto C, Van den Broeck W, Barrat A, Colizza V, Pinton J-F, Vespignani A (2010) Dynamics of person-to-person interactions from distributed RFID sensor networks. PLoS ONE 5:e11596
Cho E, Myers S, Leskovec J (2011) Friendship and mobility: user movement in location-based social networks. In: KDD '11 Proceedings of the 17th ACM SIGKDD international conference on knowledge discovery and data mining, pp 1082–1090.
Christakis N, Fowler J (2007) The spread of obesity in a large social network over 32 years. N Engl J Med 357:370–379
Christakis NA, Fowler J (2008) The collective dynamics of smoking in a large social network. N Engl J Med 358:2249–2258
Clauset A, Shalizi C, Newman MEJ (2009) Power-law distributions in empirical data. SIAM Review 51:661–703
Colizza V, Barrat A, Barthélemy M, Vespignani A (2006) The role of the airline transportation network in the prediction and predictability of global epidemics. Proc Natl Acad Sci U S A 103:2015–2020
Costa LDF, Rodrigues FA, Travieso G (2007) Characterization of complex networks: a survey of measurements. Adv Phys 56:167–242
Crandall D, Cosley D, Huttenlocher D, Kleinberg J, Suri S (2008) Feed- back effects between similarity and social influence in online communities. In KDD 08: proceeding of the 14th ACM

SIGKDD international conference on knowledge discovery and data mining New York, NY, USA, pp 160–168.
Cranshaw J, Toch E, Hong JI, Kitturr A, Sadeh N (2010) Bridging the gap between physical location and online social networks. In: 12th ACM international conference on ubiquitous, computing, pp 119–128.
Csermely P (2004) Strongs links are important, but weak links stabilize them. Trends Biochem Sci 29:331
Danon L, Arenas A, Díaz-Guilera A (2008) Impact of community structure on information transfer. Phys Rev E 77:036103
De Choudhury M, Mason WA, Hofman J, Watts DJ (2010) Inferring relevant social networks from interpersonal communication. In: WWW '10 Proceedings of the 19th international conference on, World Wide Web, pp 301–310.
Dodds P, Muhamad R, Watts DJ (2003) An experimental study of search in global social networks. Science 301:827–829
Dodds P, Watts DJ (2003) Information exchange and the robustness of organizational networks. Proc Natl Acad Sci U S A 100:12516–12521
Dorogovtesev S, Mendes JFF (2001) Effect of the accelerating growth of communication networks on their structure. Phys Rev E 63:025101
Dorogovtsev SN, Mendes JFF, Samukhin AN (2000) Structure of growing networks with preferential linking. Phys Rev Lett 85:4633–4636
Dorogovtsev SN, Mendes JFF (2002) Evolution of networks. Adv Phys 51:1079–1187
Dunbar R (1992) Neocortex size as a constraint on group size in primated. J Hum Evol 22:469
Eagle N, Pentland A, Lazer D (2009) Inferring friendship network structure by using mobile phone data. Proc Natl Acad Sci U S A 106(36):15274–15278
Eagle N, Macy M, Claxton R (2010) Network diversity and economic development. Science 328:5981
Ebel H, Mielsch L, Bornholdt S (2002) Scale-free topology of e-mail networks. Phys Rev E 66:035103
Ebel H, Davidsen J, Bornholdt S (2002) Dynamics of social networks. Complexity 8:24–27
Eckmann J-P, Moses E, Sergi D (2004) Entropy of dialogues creates coherent structures in e-mail traffic. Proc Natl Acad Sci U S A 40:14333–14337
Erdös P, Rényi A (1959) On random graphs I. Publ Math Debrecen 6:290–297
Erdös P (1960) On the evolution of random graph. Publ Math Inst Hungarian Acad 5:17–61
Everitt B (1974) Cluster analysis. John Wiley, New York
Faloutsos M, Faloutsos P (1999) On relationships of the internet topology. In SIGCOMM Comput Com 29:251–262
Feld S (1991) Why your friends have more friends than you do. Am J Sociol 95:1464–1477
Ferrara E (2011) A large-scale community structure analysis in Facebook. arXiv:1106.2503v3.
Ferrara E (2012) Topological features of online, social networks. arXiv:1202.0331v1.
Friedkin N, Johnsen E (1990) Social influence and opinions. J Math Sociol 15:193–206
Gaito S, Zignani M, Rossi G, Sala A, Wang X, Zheng H, Zhao B (2012) First ACM International Workshop on Hot Topics on Interdisciplinary Social Networks Research. Beijing, China
Gilbert E, Karahalios K (2009) Predicting tie strength with social media. In: CHI '09 Proceedings of the 27th international conference on human factors in, computing systems, pp 211–220.
Girvan M, Newman MEJ (2002) Community structure in social and biological networks. Proc Natl Acad Sci U S A 99:8271–8276
Goh K-I, Barabási A-L (2008) Burstiness and memory in complex systems. Europhys Lett 81:48002
Goh K-I, Kahng B, Kim D (2001) Universal behavior of load distribution in scale-free networks. Phys Rev Lett 87:278701
Goh K-I, Kahng B, Kim D (2002) Fluctuation-driven dynamics of the internet topology. Phys Rev E 88:108701

Golder SA, Wilkinson D, Huberman BA (2007) Rhythms of social interaction: Messaging within a massive online network. In , Steinfield C, Pentland B, Ackerman M, Contractor N (eds) Proceedings of third international conference on communities and technologies , pp 41–66
Gómez-Gardeñes J, Moreno Y (2004) Local versus global knowledge in the Barabasi-Albert scale-free network model. Phys Rev E 69:37103
Gonçalves B, Perra N, Vespignani A (2011) Modeling users' activity on Twitter networks: validation of Dunbar's number. PLoS ONE 6(8):e22656
González M, Hidalgo C, Barabási A-L (2008) Understanding individual human mobility patterns. Nature 453(7196):779–782
Grabowicz P, Ramasco J, Moro E, Pujol J, Eguiluz V (2012) Social features of online networks: the strength of intermediary ties in online social media. PLoS ONE 7(1):e29358
Granovetter M (1973) The strength of weak ties. Am J Sociol 78:1360–1380
Greene J (1997) Production and inventory control handbook, Tercera edn. McGraw-Hill, New York
Guare J (1990) Six degree of separation: a play. Random House, New York
Guimerá R, Danon L, Díaz-Guilera A, Giralt F, Arenas A (2003) Self-similar community structure in organisations. Phys Rev E 68:065103
Guimerá R, Mossa S, Turtschi A, Amaral L (2005) The worldwide air transportation network: anomalous centrality, community structure, and cities' global roles. Proc Natl Acad Sci U S A 102:7794
Guimerá R, Uzzi B, Spiro J, Amaral LA (2005) Team assembly mechanisms determine collaboration network structure and team performance. Science 308(5722):697–702
Guimerá R, Danon L, Díaz-Guilera A, Giralt F, Arenas A (2006) The real communication network behind the formal chart: community structure in organizations. J Econ Behav Organ 61:653–667
Haight FA (1967) Handbook of the poisson distribution. Wiley, New York
Hidalgo C, Rodriguez-Sickert C (2008) The dynamics of a mobile phone network. Phys A 387:3017
Holme P, Kim BJ, Yoon CN, Han SK (2002) Attack vulnerability of complex networks. Phys Rev E 65:056109
Holme P (2002) Edge overload breakdown in evolving networks. Phys Rev E 66:036119
Holme P, Edling C, Liljeros F (2004) Structure and time-evolution of an internet dating community. Soc Netw 26:155–174
Holme P (2005) Network reachability of real-world contact sequences. Phys Rev E 71:046119
Holme P, Saramäki J (2012) Temporal networks. Phys Rep 519:97–125
Huberman BA, Adamic L (1999) Internet-growth dynamics of the World-Wide Web. Nature 401:131
Huberman B, Romero D, Wu F (2009) Social networks that matter: Twitter under the microscope. First Monday 14(1):1–9
Ibarra H (2002) Homophily differential returns: sex differences in network structure and access in an advertising firm. Adm Sci Q 37:422
Iribarren J, Moro E (2009) Impact of human activity patterns on the dynamics of information diffusion. Phys Rev Lett 103:038702
Iribarren J, Moro E (2011a) Affinity paths and information diffusion in social networks. Soc Netw 33:134–142
Iribarren J, Moro E (2011b) Branching dynamics of viral information spreading. Phys Rev E 84:046116
Isella L, Stehlé J, Barrat A, Cattuto C, Pinton J-F, Van den Broeck W (2011) What's in a crowd? Analysis of face-to-face behavioral networks. J Theor Biol 166:166–180
Jeong H, Tombor B, Albert R, Oltvai Z-N, Barabási A-L (2000) The large-scale organization of metabolic networks. Nature 407:651–655
Jo H-H, Karsai M, Kertész J, Kaski K (2012) Circadian pattern and burstiness in mobile phone communication. New J Phys 14:013055
Karsai M, Kivelä M, Pan R, Kaski K, Kertész J, Barabási A-L (2011) Small but slow world: how network topology and burstiness slow down spreading. Phys Rev E 83:025102(R).
Kempe D, Kleinberg J, Kumar A (2002) Connectivity and inference problems for temporal networks. J Comput Syst Sci 64(4):820–842

Kivran-Swaine F, Govindan P, Naaman M (2011) The impact of network structure on breaking ties in online social networks: unfollowing on twitter. In: CHI '11 Proceedings of the 2011 annual conference on human factors in computing systems.
Kleinberg J (2000) Navigation in a small world. Nature 406:845
Kleinberg J (2008) The convergence of social and technological networks. Commun ACM 51(11):66–72
Korte C, Milgram S (1970) Acquaintance linking between white and negro populations: application of the small world problem. J Pers Soc Psychol 15:101–118
Kossinets G, Watts DJ (2006) Empirical analysis of an evolving social network. Science 311:5757
Kossinets G, Kleinberg J, Watts DJ (2008) The structure of information pathways in a social communication network. In: Proceedings of ACM SIGKDD '08, pp 435–443.
Kostakos V (2009) Temporal graphs. Physica A 388(6):1007–1023
Kovanen L, Karsai M, Kaski K, Kertész J, Saramäki J (2011) Temporal motifs in time-dependent networks. J Stat Mech: Theory Exp 11:11
Kumpula J, Onnela J-P, Saramäki J, Kertész J, Kaski K (2009) Model of community emergence in weighted networks. Comp Phys Comm 180:517
Kwak H, Lee C, Park H, Moon S (2010) What is Twitter, a social network or a news media? In: Proceedings of the 19th international conference on, World wide web, pp 591–600.
Lambiotte R, Blondel V, de Kerchove C, Huens E, Prieur C, Smoreda Z, Van Dooren P (2008) Geographical dispersal of mobile communication networks. Phys A 387:5317–5325
Lancichinetti A, Kivelä M, Saramäki J, Fortunato S (2010a) Characterizing the community structure of complex networks. PLoS ONE 5(8):e11976
Lancichinetti A, Radicchi F, Ramasco J, Fortunato S (2010b) Finding statistically significant communities in networks. PLoS ONE 6:e18961
Latora V, Marchiori M (2001) Efficient behavior of small-world networks. Phys Rev Lett 87:198701
Latora V, Marchiori M (2003) Economic small-world behavior in weighted networks. Eur Phys J B 32:249
Lazarsfeld P, Merton R (1954) Friendship as a Social Process: A Substantive and Methodological Analysis. In: Berger M, Abel T, Page CH (eds) Freedom and control in modern society. Van Nostrand, New York, pp 18–66
Lazer D, Pentland A, Adamic L, Aral S, Barabási A-L, Brewer N, Christakis NA, Contractor N, Fowler J, Gutmann M, Jebara T, King G, Macy M, Roy D, Van Alsyne M (2009) Computational social science. Science 323:721–723
Lee C, Scherngell T, Barber MJ (2009) Real-world separation effects in an online social network. Soc Netw 33:2
Lee S-H, Kim P-J, Ahn Y-Y, Jeong H (2010) Googling social interactions: web search engine based social network construction. PLoS ONE 5(7):e11233
Leskovec J, Lang KJ, Dasgupta A, Mahoney M (2009) Community structure in large networks: natural cluster sizes and the absence of large well-defined clusters. Internet Math 61(1):29–123
Lewis K, Kaufman J, Gonzales M, Wimmer A, Christakis NA (2008) Tastes, ties, and time: a new social network dataset using Facebook.com. Soc Netw 30:330–342
Liben-Nowell D, Novak J, Kumar R, Raghavan P, Tomkins A (2005) Geographic routing in social networks. Proc Natl Acad Sci U S A 102:623–628
Malmgren RD, Stouffer DB, Motter AE, Amaral LAN (2008) A poissonian explanation for heavy tails in e-mail communication. Proc Natl Acad Sci U S A 105:18153–18158
Mangan S (2003) Structure and function of the feed-forward loop network motif. Proc Natl Acad Sci U S A 100:11980–11985
Martin JL, Yeung K-T (2006) Persistence of close personal ties over a 12-year period. Soc Netw 28:331–362
McDaid A, Hurley N (2010) Detecting highly overlapping communities with model-based overlapping seed expansion. In: Proceedings of the ASONAM10.
McPherson J, Smith-Lovin L, Cook J (2001) Birds of a feather: homophily in social networks. Ann Rev Sociol 27:415–444

Milgram S (1967) The small world problem. Psychol Today 1:60–67
Milo R, Shen-Orr S, Itzkovitz S, Kashan N, Chklovskii D, Alon U (2002) Network motifs: simple building blocks of complex networks. Science 298:824
Miritello G, Lara R, Cebrián M, Moro E (2012a) Limited communication capacity unveils strategies for human interaction. arXiv:1304.1979.
Miritello G, Moro E, Lara R (2011) Dynamical strength of social ties in information spreading. Phys Rev E 83:045102(R).
Miritello G, Moro E, Lara R, Martinez R, Belchamber J, Roberts S, Dunbar R (2012b) Time as a limited resource: communication strategy in mobile phone networks. Social Networks 35:89–95.
Mislove A, Marcon M, Gummadi K, Druschel P, Bhattacharjee B (2007) Measurement and analysis of online social networks. In: Proceedings of the 7th ACM conference on internet, measurement, pp 29–42.
Mollica K, Gray B, Trevino L (2003) Racial homophily and its persistence in newcomers' social networks. Organ Sci 14:123–136
Moody J (2002) The importance of relationship timing for diffusion. Soc Forces 81:25–56
Moreno Y, Pastor-Satorras R, Vespignani A (2002) Epidemic outbreaks in complex heterogeneous networks. Eur Phys J B 26:521–529
Nanavati AA, Singh R, Chakraborty D, Dasgupta K, Mukherjea S, Das G, Gurumurthy S, Joshi A (2008) Analyzing the structure and evolution of massive telecom graphs. IEEE Trans Knowl Data Eng 20:5
Newman MEJ (2001a) Scientific collaboration networks: I. Network construction and fundamental results. Phys Rev E 64:016131
Newman MEJ (2002b) The spread of epidemic disease on networks. Phys Rev E 66:016128
Newman MEJ, Forrest S, Balthrop J (2002) Email networks and the spread of computer viruses. Phys Rev E 66:035101
Newman MEJ (2002a) Assortative mixing in networks. Phys Rev E 89:208701
Newman MEJ (2003a) Properties of highly clustered networks. Phys Rev E 68:026126
Newman MEJ, Park J (2003) Why social networks are different from other types of networks. Phys Rev E 68:036122
Newman MEJ (2003b) The structure and function of complex networks. SIAM Review 45:167–256
Newman MEJ, Girvan M (2004) Finding and evaluating community structure in networks. Phys Rev E 69:026113
Oliveira J, Barabási A-L (2005) Human dynamics: Darwin and Einstein correspondence patterns. Nature 437:1251
Onnela J, Chakraborti A, Kaski K, Kertész J, Kanto A (2003) Dynamics of market correlations: taxonomy and portfolio analysis. Phys Rev E 68:056110
Onnela J, Saramäki J, Kertész J, Kaski K (2005) Intensity and coherence of motifs in weighted complex networks. Phys Rev E 71:065103
Onnela J-P, Saramäki J, Hyvönen J, Szabó Z, Argollo de Menezes M, Kaski K, Barabási A-L, Kertész J (2007a) Analysis of a large-scale weighted network of one-to-one human communication. New J Phys 9:179
Onnela J-P, Saramäki J, Hyvönen J, Szabó Z, Lazer D, Kaski K, Kertész J, Barabási A-L (2007b) Structure and tie strengths in mobile communication networks. Proc Natl Acad Sci U S A 104:7332
Onnela J-P, Arbesman S, González M, Barabási A-L, Christakis NA (2011) Geographic constraints on social network groups. PLoS ONE 6(4):e16939
Palla G, Barabási A-L, Tamás V (2007a) Community dynamics in social networks. Noise Stochast Complex Syst Finan 6601:660106
Palla G, Barabási A-L, Vicsek T (2007b) Quantifying social group evolution. Nature 446:664–667
Pan R, Saramäki J (2011) Path lengths, correlations, and centrality in temporal networks. Phys Rev E 84:016105
Pastor-Satorras R, Vespignani A (2001b) Epidemic spreading in scale-free networks. Phys Rev Lett 86:3200–3203
Pastor-Satorras R, Vespignani A (2002) Immunization of complex networks. Phys Rev E 65:036104

Pastor-Satorras R, Vespignani A (2004) Evolution and structure of the internet: a statistical physics approach. Cambridge University Press, Cambridge

Quattrocchi W, Conte R, Lodi E (2010) Simulating opinion dynamics in heterogeneous communication systems. In: Proceedings of 7th european conference on complex systems (ECCS), pp 70–85.

Raeder T, Lizardo O, Chawla N, Hachen D (2011) Predictors of short-term decay of cell phone contacts in a large scale communication network. Soc Netw 33:245–257

Raghavan U, Albert R, Kumara S (2007) Near linear time algorithm to detect community structures in large-scale networks. Phys Rev E 76:036106

Reynolds P (2003) Call center staffing. The Call Center School Press, London

Roberts S (2010) Constraints on social networks. Proc Br Acad 158:115–134

Rocha L, Liljeros F, Holme P (2010) Information dynamics shape the sexual networks of internet-mediated prostitution. Proc Natl Acad Sci U S A 107:5706–5711

Romero D, Meeder B, Barash V, Kleinberg J (2011) Maintaining ties on social media sites: the competing effect of balance, exchange and betweenness. In: Proceedings of the ICWSM'11, fifth international AAAI conference on weblogs and social media.

Rosvall M, Bergstrom C (2008) Maps of information flow reveal community structure in complex networks. Proc Natl Acad Sci U S A 105:1118–1123

Rybski D., Buldyrev SV, Havlin S, Liljeros F, Makse HA (2010) Communication activity: temporal correlations, clustering, and, growth. arXiv:1002.0216v1.

Rybski D, Buldyrev SV, Havlin S, Liljeros F, Makse HA (2009) Scaling laws of human interaction activity. Proc Natl Acad Sci U S A 106:12640

Saramäki J, Kivelä M, Onnela J-P, Kaski K, Kertész J (2007) Generalizations of the clustering coefficient to weighted complex networks. Phys Rev E 75:027105

Sarkar P, Moore A (2005) Dynamic social network analysis using latent space models, Special edn. Link Mining, In SIGKDD Explorations

Scellato S, Noulas A, Lambiotte R, Mascolo C (2011) Socio-spatial properties of online location-based social networks. In: ICWSM'11, Fifth international AAAI conference on weblogs and social media.

Scott J (2000) Social network analysis: a handbook. Sage Publications, London

Snijders, Tom AB, The Statistical Evaluation of Social Network Dynamics. pp. 361–395 in Sociological Methodology (2001), edited by M.E. Sobel and M.P. Becker. Boston and London: Basil Blackwell.

Solé RV, Pastor-Satorras R, Smith E, Kepler TB (2002) A model of large-scale proteome evolution. Adv Complex Syst 5:43

Sporns O (2003) Network analysis, complexity, and brain function. Complexity 8:56

Strogatz SH (2001) Exploring complex networks. Nature 410:268–276

Szabó G, Barabási A-L (2006) Network effects in service usage.arXiv:physics/0611177.

Tang J, Musolesi M, Mascolo C, Latora V (2010) Characterising temporal distance and reachability in mobile and online social networks. ACM SIGCOMM Comput Commun Rev 40:1

Tang J, Musolesi M, Mascolo C, Latora V (2009) Temporal distance metrics for social network analysis. In: Proceedings of the 2nd ACM SIGCOMM workshop on online social networks (WOSN'09), Barcelona, Spain.

Tang W, Zhuang H, Tang J (2011) Learning to infer social ties in large networks. In: Proc. ECML PKDD'11. Proceedings of the 2011 european conference on machine learning and knowledge discovery in databases, vol Part III. pp 381–397.

Tantipathananand C, Berger-Wolf T, Kempe D (2007) A framework for community identification in dynamic social networks. In: Proc. KDD '07 Proceedings of the 13th ACM SIGKDD international conference on knowledge discovery and data mining, pp 717–726.

Toivonen R, Onnela J-P, Saramäki J, Hyvönen J, Kaski K (2006) A model for social networks. Physica A 371:851

Toivonen R, Kumpula J, Saramäki J, Onnela J-P, Kertész J, Kaski K (2007) The role of edge weights in social networks: modelling structure and dynamics. In: Proceedings of SPIE: noise and stochastics in complex systems and finance, Proc. of SPIE 6601, 660110.
Ugander J, Karrer B, Backstrom L, Marlow C (2011) The anatomy of the facebook social graph. preprint:arXiv:1111.4503v1.
Valverde S, Solé RV (2005) Network motifs in computational graphs: a case study in software architecture. Phys Rev E 72:026107
Vázquez A, Flammini A, Maritan A, Vespignani A (2003) Modeling of protein interaction networks. ComPlexUs 1:38
Vázquez A, Dobrin R, Sergi D, Eckmann J-P, Oltvai Z-N, Barabási A-L (2004) The topological relationship between the large-scale attributes and local interaction patterns of complex networks. Proc Natl Acad Sci U S A 101:17940–17945
Vázquez A, Oliveira J, Dezsö Z, Goh K-I, Kondor I, Barabási A-L (2006) Modeling bursts and heavy tails in human dynamics. Phys Rev E 73:036127
Vázquez A, Rácz B, Lukács A, Barabási A-L (2007) Impact of non-Poissonian activity patterns on spreading processes. Phys Rev Lett 98:158702
Volkovich Y, Scellato S, Laniado D, Mascolo C, Kaltenbrunner A (2012) The length of bridge ties: structural and geographic properties of online social interactions. Proceedings of the sixth international AAAI conference on weblogs and social media, In
Waleaevens J, Demoor T, Maertens T, Bruneel H (2012) Stochastic queuing-theory approach to human dynamics. Phys Rev E 85:021139
Wasserman S, Faust K (1994) Social networks analysis. Cambridge University Press, Cambridge
Watts DJ, Strogatz SH (1998) Collective dynamics of 'small-world' networks. Nature 393:440–442
Watts DJ (1999) Small worlds: the dynamics of networks between order and randomness. Princeton University Press, Princeton, NJ
Watts DJ (2004) The "new" science of networks. Annu Rev Sociol 30:243–270
Wellman B, Wortley S (1990) Different strokes from different folks: community ties and social support. Am J Sociol 96:558–588
Wellman B, Salaff J, Dimitrova L, Garton L, Gulia M, Haythornthwaite C (1996) Computer networks as social networks: collaborative work, telework, and virtual community. Annu Rev Sociol 22:213
Wellman B (2001) Computer networks as social networks. Science 293:2031
Wellman B (2007) The network is personal: introduction to a special issue of social networks. Soc Netw 29(3):349–356
Wellman B, Haythornthwaite C (2003) The internet in everyday life. Blackwell, Oxford
West D (1995) Introduction to graph theory. Prentice-Hall, Englewood Cliffs, NJ
Wilkinson D, Huberman B (2004) A method for finding communities of related genes. Proc Natl Acad Sci U S A 101:5241–5248
Wu T, Zhoud C, Xiaob J, Kurthsa J, Schellnhubera H (2010) Evidence for a bimodal distribution in human communication. Proc Natl Acad Sci U S A 107:18803
Wuchty S (2009) What is a social tie? Proc Natl Acad Sci U S A 106:15099–15100
Wuchty S, Uzzi B (2011) Human communication dynamics in digital footsteps: a study of the agreement between self-reported ties and email networks. PLoS ONE 6(11):e26972
Yook S, Jeong J, Barabási A-L, Tu Y (2001) Weighted evolving networks. Phys Rev Lett 86:5835
Zhao Q, Oliver N (2010) Communication motifs: a novel approach to characterize mobile communications. In: NetMob2010.

Chapter 3
Social Strategies in Communication Networks

For most of us time is a scarce resource. In our daily lives we are constantly allocating time among various activities and people.
—Christopher Winship

Within the wide spectrum of activities in which humans participate on a daily basis, such as work, sleep or engage in entertainment and sport activities, also social interaction can be seen as a task to be executed that, as well as any other social assignment, is constrained by the fixed amount of resources and time people have. Traditionally, the study of how people allocate their time among different activities, such as work or leisure, has been a subject of economic studies (Becker 1965; Ghez and Becker 1975; Linder 1960). Here, however, we are interested in the analysis of how of how people allocate time and attention among one another, a topic that has been usually addressed from sociological and psychological perspectives (Dunbar 1992; Granovetter 1973; Southerton 1986; Wellman 2007; Winship 1978), aimed to understand how this dynamics might affect phenomena such as network cohesion (Ho et al. 2006; Gargiulo and Benassi 2000), strategy and cooperation (Burt 1992; Seibert et al. 2001) and organizational performance (Benner and Tushman 1999; March 1991).

Any type of social interaction requires indeed social commitment and time investment, from meeting somebody face-to-face to sending an e-mail or making a phone call. As a consequence, although on one hand building and maintaining social relationships may provide social benefits and profits (De Graff and Flap 1988), on the other hand it is a resource-consuming task (Hansen et al. 2001) and requires, beside obligations, expectations or social norms (Coleman 1988b), also dedication, time and attention. Since people only have a limited amount of time per day and only a fraction can be dedicated to social interactions, we expect that they follow some strategy of time allocation to maintain their close contacts and/or establish new ones.

As we have seen in Sect. 2.2.2, peoples' capacity to maintain social contact is limited (Dunbar 1992; Roberts 2010) and despite the large number of relationships

"The Allocation of Time Among Individuals" (Winship 1978).

G. Miritello, *Temporal Patterns of Communication in Social Networks*, Springer Theses, DOI: 10.1007/978-3-319-00110-4_3,

an individual may establish in a given period or during his lifetime, this number reduces significantly when considering only those people he actually interacts with (Ferrara 2011; Gonçalves et al. 2011; Grabowicz et al. 2012; Huberman et al. 2009; Leskovec et al. 2009; Marlow 2009; Saramäki et al. 2012).

Recently, the increasing availability of massive empirical data sets of human communication has offered huge opportunities to validate sociological theories and uncover the mechanisms governing time allocation in social networks (Gonçalves et al. 2011; Miritello et al. 2012; Onnela et al. 2007; Saramäki et al. 2012). What emerge is that time constraints appear in particular on those individuals with large social circles who actually, only interact with a fraction of their connections (Ferrara 2011; Grabowicz et al. 2012; Huberman et al. 2009; Leskovec et al. 2009; Marlow 2009) and tend to dedicate, on average, less time to each of them than people who have small social circles (Gonçalves et al. 2011; Miritello et al. 2012). These results are in line with Dunbar's theory (Dunbar 1992), which asserts that the time people can dedicate to social relationship is limited by cognitive and biological constraints (see Sect. 2.2.2). In addition, within an individual's network not all the connections have the same strength or importance: family, friends, work colleagues, acquaintances which, of course, also reflects in the time people dedicate to each of them: strong ties take more time than weak ties (Granovetter 1973; Roberts 2010; Wellman and Wortley 1990). Beside the importance that it may have from a sociological point of view, the analysis of how people schedule tasks and allocate time has important implications also on the modeling of human related activities such as opinion dynamics (Friedkin and Johnsen 1990; Quattrocchi et al. 2010), trust and influence (Coleman 1990; Friedkin and Johnsen 1990; Putnam 1993) and information spreading (Aral and Van Alsyne 2007; Bakshy and Rosenn 2012; Burt 1992), since all these processes depend on the way people interact and the time they dedicate to their connections.

The aim of this chapter is to understand how people schedule and manage their attention across their network, whether different social strategies of communication exist and how they relate to both structural and temporal properties of human interactions. The problem of time allocation has been usually addressed by considering a static picture of the contact network, in which the number of contacts one individual can manage has been inferred from the aggregated number of revealed connections within the observation period, as well as his activity (see Sect. 2.3). The latter constitutes the starting point of our study. By analyzing the aggregated social network resulting from mobile communication over a long time period, we first examine what we call *static social strategies* of communication. In particular, we investigate whether the time people dedicate to social interaction is related to the size of their network and analyze the diversity in the way they distribute such a time. In line with previous studies (Dunbar 1992; Gonçalves et al. 2011), we observe that people with large social networks are more affected by time constraints than people with few connections. We also show that there are some universal aspects in the way users organize their time/attention across their network that, instead, go beyond the differences in their topological features (Miritello et al. 2012).

This analysis, however, neglects the fact that social relationships have very different activation/deactivation times (Gaito et al. 2012; Leskovec et al. 2008) and, as we will see in Chap. 4, they are characterized by very different lifetimes (Miritello et al. 2013). This means that, when we focus on a certain time window, the connections of a given individual are not active all the time, while they form and decay at a very high pace (Hidalgo and Rodriguez-Sickert 2008; Raeder et al. 2011). As a consequence, the aggregate social connectivity does not capture the instantaneous contact network an individual interacts with and it may only constitute a upper limit to the real social capacity. The same applies to the time people dedicate to social relationships. Since human activity is very bursty in time (Barabási 2010), the mere number of interaction events across a given time window does not provide any information on the way in which the individual attention is distributed across his network. As discussed in Sect. 2.4, in recent years much attention has been paid to understand both the dynamics of tie creation and decay and the mechanisms responsible for the bursty behavior. The latter, in particular, has been also related to priority queues in the way in which individuals execute tasks (Barabási 2005; Vázquez et al. 2006). However, what has not been thoroughly investigated is the impact that temporal dynamics have on the way in which people interact with their network. Indeed, these temporal inhomogeneities continuously alter the structure of the networks in which each individuals participates, leading to ongoing changes in the instantaneous contact network of an individual and make it very different from the aggregated one (Palla et al. 2007; Tantipathananand et al. 2007). It is therefore plausible to expect that they also play a crucial role in the individual strategies of communication and time allocation.

To address this we then study the instantaneous, instead than the aggregated, mobile phone network, which allows us to characterize what we call dynamical social strategies what we call *dynamical social strategies* (Miritello et al. 2012). Contrary to the static network, the study of the instantaneous and dynamic one has not received much attention in the analysis of social networks; this is due to the fact that, despite its interest, it is a very difficult and challenging problem.

The main difficulty when analyzing the instantaneous interaction network is probably due to the fact that there is no explicit rule to assess tie formation and decay. Although in some online social networks there are explicit rules for establishment of social ties, in most communication channels like mobile-phone calls or e-mail, the only way to assess the existence or not of a tie is by looking at the tie activity. In addition, accordingly to traditional approaches, tie formation/decay is considered a much slower process than tie interaction. For these reasons, in most studies, a tie is assumed to be present if it shows any kind of activity in the observation window (Hidalgo and Rodriguez-Sickert 2008; Onnela et al. 2007). However, since communication is bursty (Barabási 2010), large inter-event times are likely and interactions might be unobserved or mistaken as decay or formation of ties, especially if the observation window is short. In our database, for example, we find that the average time between tie communication events is 14 days. Thus, if the observation period is of the order of months, we might get apparent non-trivial growth of the social connectivity as ties are revealed in time which would lead to an erroneous determination of an individual's network.

In our analysis we avoid the latter problem by studying the mobile phone network during a very long (almost 2 years) data set. Moreover, to infer tie formation/decay from activity data we develop a criterium to disentangle ties communication from their evolution. This method is actually the key to investigate dynamically the process of time allocation, since it allows us to analyze the instantaneous social network of a given individual. This analysis leads us to important findings. We observe that, contrary to the perception of ever-growing social connectivity, people exhibit a *social capacity*, which limits the number of ties they can maintain opened during a certain time period. Social capacity, that can be interpreted as the instantaneous social connectivity, can be significantly different from the aggregated number of social relationships. On the other hand, we use the information on the number of ties hat people form and destroy in time to define the individual *social activity*. We are therefore able to distinguish, within all the possible connections of a given individual, between those maintained over time and more volatile ones. Interestingly, we observe that people manage these two types of connections in such a way that, despite the turnover, their social capacity remains almost constant in time. Furthermore, multiple combinations of social capacity and activity reveal a diverse range of social strategies, from exploratory to stable. Such strategies, which can not be captured from the aggregated social connectivity, change with the age and gender of the individual and are highly assortative. The assortative nature of dynamical communication strategies have important implications on the resulting topology of the global network and in phenomena such as information diffusion (Miritello et al. 2012).

Our analysis goes far beyond the characterization of time allocation strategies in social networks and, actually, it provides general results and methods to improve the current description of social networks, from aggregated (or ever-growing) objects to evolutionary quantities.

3.1 Static Social Strategies

As mentioned above, we first investigate the static strategies of time allocation and study the way in which the time people dedicate to social relationships is related to the size of their personal network (Miritello et al. 2012). To this end, from the set of Call Detail Records (CDRs) described in Appendix A, we first analyze the aggregated weighted communication network over a period T of 11 months, where the weight (or intensity) $w_{ij} = w_{ji}$ of a tie $i \leftrightarrow j$ is the aggregated duration of calls between users i and j during the whole time window. As ahown in Appendix A, considering the total duration or the total number of phone calls between two individuals to define tie weight, leads to an equivalent description of the contact network (see Fig. A.2). As a consequence, the choice of one or the other quantity basically depends on the case under investigation. Since here we are interested in analyzing the way in which people distribute their limited time across their social relationships, it makes more

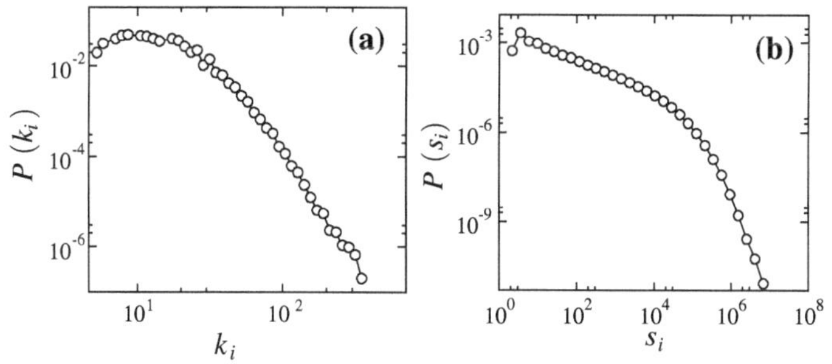

Fig. 3.1 Distribution of the social connectivity k_i (**a**) and the nodes' strength (**b**) of the mobile phone network under consideration. Aa large heterogeneity in both quantities is observed

sense to define w_{ij} as the aggregated duration of phone communications. As we will see in Chap. 5, in situations where instead the duration of the interaction does not constitute an essential ingredient, the total number of phone calls may be a more convenient choice.

3.1.1 The Boundaries of Human Communication

Before analyzing how individuals' cognitive and time constraints reflect in the distribution of attention across their social relationships, it is important to characterize the global features of the network. In general, we find that the topology of our network exhibits properties similar to the ones observed in previous studies of mobile communication networks (Onnela et al. 2007). Indeed, as shown in Fig. 3.1, we found a skewed distribution with a fat tail for both the nodes' social connectivity k_i and strength s_i. In our database, we observe that the mean of social connectivity is around 85, with a maximum value of around 400. For the node strength s_i instead, we found that, although the mean of this distribution is around 1.5 h in the whole period, the maximum value is about 6 h per day. This means that, while the time that the larger part of the population spends on the phone per day is of the order of seconds or minutes, there is a small minority who phone more than 1 h per day. Not only the aggregated s_i, but also the ties weight w_{ij} show a long-tailed distribution, which indicates a strong heterogeneity in the way people distribute the time across their social circle. For both k_i and s_i, however, the decay is faster than a power-law, which indicates the presence of a relatively small number of hubs. As mentioned in Sect. 2.2, this decay is probably due to the fact that we have filtered out business phone numbers, which mainly correspond to the hubs in mobile networks, a possibility also pointed out by Onnela et al. (Onnela et al. 2007). The fact that the functional behavior of the distribution $P(s_i)$ exhibits similarities with the degree distribution $P(k_i)$ is not unexpected since in many communication networks the strength of a node increases with its degree, thus the slow decaying tail of $P(s_i)$

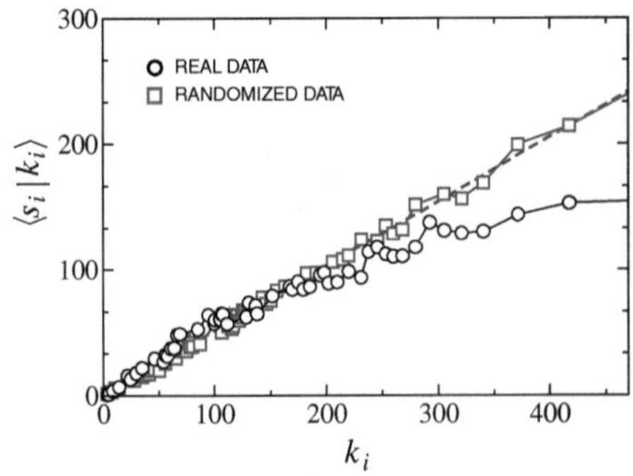

Fig. 3.2 Average strength s_i of a node (measured as the aggregated duration, in hours, of phone calls) for a given social connectivity as a function of the social connectivity k_i. The *open circles* correspond to the real data and the *open squares* to the randomized network. For real data the strength of nodes saturates for relatively large values of connectivity, while in the randomized case the data can be fitted by the uncorrelated approximation $s_i = \langle w \rangle \, ki$, corresponding to the *blue dashed line*

reflects the decay of $P(k_i)$. To shed more light on the dependence of the total strength of a node on the size of its social circle, we first analyze the relationship between these two quantities. To assess significance to our observations, we also compare the results with a randomized network, where tie weights are randomly chosen among the whole population of ties (see Appendix A). Note that in the randomized network the overall social connectivity of each user (thus the network topology) is preserved, while the amount of time each user dedicates to all his connections does not now correspond to the actual value. Note also that in this analysis we concentrate on the egocentric network of a given user i.

We analyze the way in which the time he spends on the phone correlates with the topology of his network, regardless to the topological properties of other users. As shown in Fig. 3.2, we observe that the strength s_i of a node increases with the total number of its connections k_i: people with many contacts invest much more time in communication than people with few of them. This result is in line with previous studies (Barrat et al. 2004), where authors find that the average strength $s(k)$ of nodes with degree k increases with k as $s(k) \sim k^\beta$ for both scientific collaboration and the air-transportation networks. Nevertheless, we observe a slightly more complex behavior. For small values of k_i, the average strength of nodes increases almost linearly with the connectivity. However, for larger values of k_i the behavior of s_i changes: it gradually starts to grow sub-linearly until it saturates for very large values of k_i, which suggests the existence of a limit in the user's ability to allocate time proportionally to their social connectivity. The latter reflects the fact that time is finite, thus if users add new contacts to their network, the time they invest to communicate with them does not necessarily increase in a proportional fashion. Indeed, as shown in Fig. 3.2 we observe a clear deviation from the linear approximation from $k_i \simeq 150$ onwards. In the same figure we also show the results obtained for the randomized network. In this case the average nodes' strength is very well fitted by the linear approximation $s_i = \langle w \rangle k_i$ which means that, given the number of connections of a node, the corresponding average strength is provided as well and the two quantities provide the same information on the system. As mentioned in Sect. 2.1.2, this linear behavior indicates that at the average level tie weights are mostly uncorrelated to the degree of the node i. In this case, in fact, we can approximate $w_{ij} = \langle w \rangle$, where $\langle w \rangle$ is the average tie weight in the whole network. The deviations from such a linear behavior observed for real data suggest the existence of correlations between s_i and k_i. In particular, the fact that s_i increases sub-linearly for large k_i indicates that, on average, highly connected people tend to spend less time on the phone than the one that they would spend with a random assignment of weights. This result shows very clearly the difference between mobile phone networks and other types of social networks. For example, for scientific collaboration and air transportation networks a linear and a super-linear behavior, respectively, has been observed (Barrat et al. 2004). While the former suggests that a larger number of collaborators leads to more publications (or the other way around), the latter indicates that, for air-transportation networks, the larger is the airport the more traffic it can handle. We observe that neither the former nor the latter conclusion can be drawn for human interactions where a larger number of contacts does not necessarily imply a larger investment of time (and money) in communication.

The existence of a limit in the communication time is better observed by looking at the average weight of ties of each user, which we refer as $w_i = \sum_{j=1}^{k_i} w_{ij}/k_i = s_i/k_i$, as a function on the social connectivity k_i, which is shown for both the real and the randomized network in Fig. 3.3. Here it becomes much more clear that, on average, the time dedicated to each connection gradually increases with k_i: the

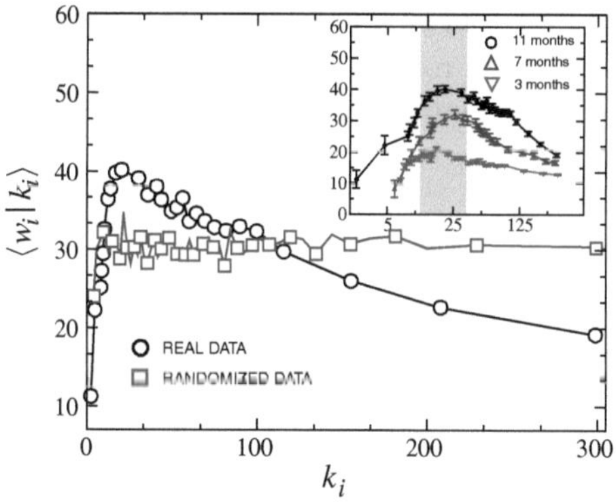

Fig. 3.3 Average weight w_{ij} of the ties (in hours) as a function of the social connectivity k_i. For the real data the average weight of each tie gradually increases with the connectivity until it reach a maximum, which is estimated to be around 20 connections. No significant dependence is observed in the randomized case. In the inset we compare the results obtained in the observation time period of 11 months (*open black circles*) with the ones obtained for a period of 3 and 7 months (*green down-side* and *blue up-side triangles* respectively)

more relationships people have, the more time they need to dedicate on average to each of them. However, when the number of connections surpasses a certain threshold, which is around $k_i \in [10, 40]$, the user can no longer dedicate the same amount of time to each of them. This is why the average value of w_{ij} reaches a maximum, then starts to decrease with k_i. The observed behavior can be related to Dunbar's theory (Dunbar 1998), discussed in Sect. 2.2.2, which asserts that cognitive and biological constraints limit the number of people an individual keeps social contact with. Interestingly, in our data, the saturation lies significantly below the Dunbar number of 150–200 connections, probably because beside the cognitive limit, also temporal and monetary constraints play their role in phone communication. This may explain the difference between our results and the number of stable relationships a user can maintain in an online social networks as Twitter, which has been found to lie around the Dunbar number of 150–200 (Gonçalves et al. 2011). Furthermore, contrary to phone communication, online social networks have a cheaper social cost since they do not require an intimate or at least face-to-face acquaintance and little work is done to maintain a large numbers of hardly known contacts. There are however other possibilities: the limit of 10–40 that we observe might reflect the maximum number of more intense relationships, which has been found to be lie 30 alters (Hill and Dunbar 2003; Sutcliffe et al. 2012), with the rest above this number being contacted only very occasionally. Moreover, some contacts are seen face-to-face rather than called on mobile phones. Finally, it should be also noted that, in common with other studies (Onnela et al. 2007; Palla et al. 2007), the personal network captured by the mobile phone records is not the complete personal network, but just those network members who are with the same phone company as the user, which may diminish the actual value of the real saturation point.

Interestingly, we find that the position of the peak does not change with the length of the time period. This is shown in the inset of Fig. 3.3, where we compare the results obtained in the time period of 11 months with the observation within 3 and 7 months. This finding demonstrates the robustness of our result and shows that the limit in people's social capacity is an intrinsic constraint and not an effect of the finite time window (Miritello et al. 2012).

3.1.2 Time Allocation Diversity

The results in Fig. 3.3 indicate that people with larger networks dedicate on average less time to each of their social connections. This raises the question on whether the heterogeneity observed in the distribution of tie weights is due to the existence of communication strategies. Due to the limited amount of time people can devote to social relationships, it is in fact plausible that they follow different strategies to allocate their time, according to the total size of their personal network or the total time they dedicate to phone communication. In this case, the observed heterogeneity of w_{ij} could be due to the existence of users with different communication strategies.

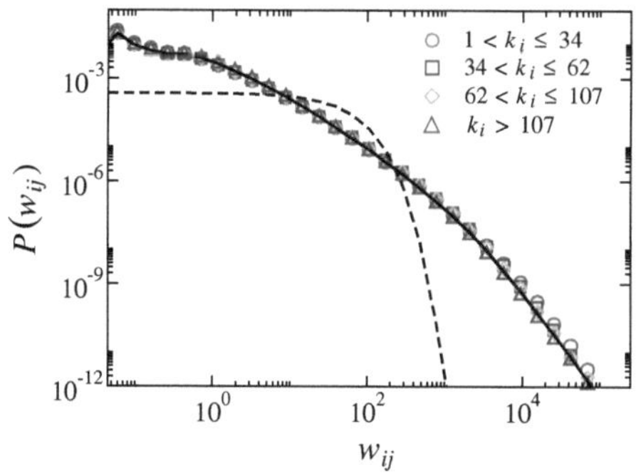

Fig. 3.4 Distribution of tie weights for nodes with different values of social connectivity that correspond to the quartile of the distribution of k: $1 < k \leq 34$ (*magenta circles*), $34 < k \leq 62$ (*blue squares*), $62 < k \leq 107$ (*green diamonds*) and $k \geq 107$ (*red triangles*) compared to the one obtained when the whole population is considered (*black solid curve*). The distribution does not show a significant dependence on the social connectivity. For comparison, we also show an exponential distribution with the same mean of the real distribution (*black dashed curve*)

Thus, for example, one might assume that users with a large number of connections contribute to small weights more than users with few connections.

To address this, we analyze the distribution of tie weights for users with different values of the social connectivity and found that, instead, the distribution of w_{ij} does not show an appreciable dependence on k_i. This result is shown in Fig. 3.4, where each curve corresponds to the distribution of tie weights for nodes belonging to different intervals of social connectivity, chosen accordingly to the quartiles of the whole distribution of k_i (only users with $k_i > 1$ are considered). Our finding indicates that people always distribute their time unevenly across their contacts: they dedicate a small amount of time to many people and a large amount of time to a small number of people, independently of the size of their social circle. To investigate users' diversity in time allocation in more detail, we measure the *disparity* Y_i, a widely used measure of diversity in the network literature which has been introduced in Sect. 2.1.2. Although the concept of disparity is not a new one (Barthélemy et al. 2003, 2005; Boccaletti et al. 2006; Lee et al. 2010), it has received relatively little attention in mobile-phone communication networks. Eagle et al. (2010) examined the diversity of individuals' relationships and its connection with the economic development of communities in which they live. However the diversity and, especially, its relationship with degree and node strength has not been thoroughly investigated for communication networks. Figure 3.5 shows $k_i Y_i$ as a function of the social connectivity k_i. When all the ties have the same strength, this quantity is $k_i Y_i = 1$ and does not depend on k_i, while if the distribution is severely heterogeneous we have $k_i Y_i = k_i$.

Our results are intermediate between the two extreme cases of perfect homogeneity and perfect heterogeneity. Specifically, we find that for $k_i > 20$ the curve is well fitted by the relation $k_i Y_i \simeq k_i^{\alpha}$ with $\alpha = 0.5$. The exponent α smaller than 1 indicates a dependence between the social connectivity and the disparity, thus suggesting

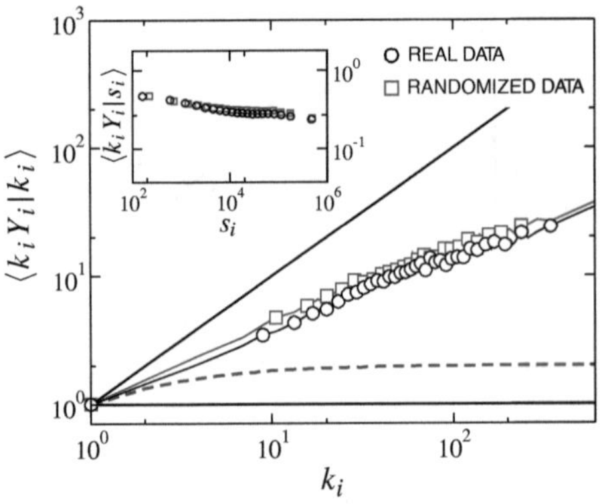

Fig. 3.5 Average $k\ Y_i$ as a function of the social connectivity k_i. No significant difference is observed between the real network (*black circles*) and the randomized one (*red squares*). Our data are intermediate between the two extreme cases of perfect homogeneity $kY = 1$ and perfect heterogeneity $kY = k$ (*solid lines*), with an exponent of 0.5 for large connectivities. The *dashed blue line* represents the case in which the weights are distributed according to an exponential distribution, which shows a fast saturation to the homogeneous case. In the inset is shown $k\ Y_i$ as a function of the node's strength s_i. No significant correlation is observed between these two quantities, a result obtained also for the randomized network

the existence of different strategies of communication between users who have large numbers of connections and those who have few. However, as shown in Fig. 3.5, we find that the same result is obtained after randomizing the weights of the ties over the whole network. For a given social connectivity, the disparity of the real case (black circles) is always slightly smaller than the one obtained in its randomized version (red squares), indicating that the real communication is slightly more homogeneous than the one corresponding to a random assignment of tie weights. Nevertheless, no significant difference is observed between the real network and the randomized one in the scaling of Y_i, suggesting that the way users organize their time/attention with each one of their contacts does not alter the diversity in communication (Miritello et al. 2012).

This is an important finding since the dependence of the disparity of the flux that passes through a node (in our case the communication time) on its degree or strength is often used to assess the existence of nodes that have different functionalities within the network (Almaas et al. 2004; De Montis et al. 2007; Lee et al. 2010). We rather observe that it is just a reflection of the long tailed nature of the distribution of the weights, which introduces a strong heterogeneity. In fact, the larger the connectivity of a node, the higher is the probability that its ties have weights belonging to the tail of the distribution. This is why the observed behavior differs from the homogeneous $kY_i = 1$ curve. To test this we calculated the disparity of nodes where the tie weights are now randomly chosen from an exponential distribution with the same mean of the real one. As shown by the dashed curve in Fig. 3.5, in this case we observe a fast saturation of Y_i to the homogeneous case in which Y_i is independent of k_i. Note that also in this case, a small deviation from the $\alpha = 1$ behavior is observed for very small values of k. Due to this small-size effect the exponent in the real data is smaller than 1, which is the reason why the curve $k_i Y_i$ saturates to a constant value for very large values of connectivity.

A plausible explanation of the observed relation between Y_i and k_i comes therefore from the heterogeneity in w_{ij} and the fact that the disparity measure Y_i is very sensible to the distribution of w_{ij}. Actually, if we define ties with small and a large weights respectively as *weak* and *strong* ties, the result in Fig. 3.4 indicates that the proportion between strong and weak ties does not change with the social connectivity. Therefore, the larger the social connectivity, the larger the number of strong contacts has to be. Despite the high correlation between the disparity and the social connectivity, the inset of Fig. 3.5 shows that the disparity is mostly independent of the strength of the node. This is due to the fact that the relation between the strength and the social connectivity is not univocal since for a given s_i there are many different values of k_i which in turn give many possible values of disparity (Miritello et al. 2012).

3.2 Social Strategies, Bursty Activity and Tie Dynamics

So far we have analyzed time allocation in human communication by considering the aggregated activity of users across the whole observation time period. However, the problem of time allocation is a dynamical process and it is plausible to expect that it is affected not only by the aggregated but also by the instantaneous size of the individuals' network and activity. Indeed, since communication events are not evenly distributed across the observation time window and social relationships are continuously added and removed in time (see Sect. 2.4), we might expect that individual dynamics of time allocation depend also on such temporal inhomogeneities. In the rest of the chapter we will analyze this latter issue. Specifically, we refer to *dynamical* strategies of communication (or dynamical social strategy) to distinguish them from the *static* or aggregated social strategies analyzed in the previous section. As mentioned above, two main reasons have hampered a deep understanding of dynamical social strategies of communication: on the one hand, the limited availability of long time data sets and, on the other hand, the high entanglement between the time scale of users' (and ties) communication activity and the one of tie formation and decay. By analyzing a large set of call records during a time period of 19 months (February 2009 to August 2010) we first show that, indeed, the bursty patterns of human communication leads to an apparent dynamics of the social connectivity over time. Then, to overcome the difficulty of the time-scale entanglement, we develop a criterium to separate tie activity from tie formation/decay. The latter approach allows us to investigate with high precision the individual dynamical communication strategies presented in the next section.

3.2.1 Apparent Dynamics of Social Connectivity

In most studies of communication networks it is assumed that a tie forms at the time when the first communication is observed. However, inferring the formation (or the decay) of a tie is hampered by the very dynamics of the interaction. As we have discussed in Chap. 2, communication within ties is bursty and inter-event times have a heavy-tailed distribution. Thus, since large inter-event times are likely, interactions might be unobserved or confused with their formation and decay if the observation window is short. For example, in our database we find that the average inter-event time is $\langle \delta t_{ij} \rangle = 14$ days while the standard deviation is around 18 days. Therefore, if the observation period is of the order of months we might get an apparent non-trivial growth of the social connectivity as ties as revealed slowly in time. According to models of network growth (Newman 2003b), the observed connectivity of a node as a function of time seems to have an increasing behavior $k(t) \sim t^{\gamma}$ with $\gamma \simeq 1/2$. Here we show that this effect, however, is (mostly) due to the fact that different ties have different average inter-event times $\overline{\delta t}_{ij}$ in the observation period or, equivalently, different numbers of communication events w_{ij}. Let us consider a tie which is observed also before and after the observation window and which inter-event time distribution is given by $P(\delta t_{ij})$. Assuming that we chose randomly the starting time of the observation window, the time to the first observation of the link is given by the waiting time (Breuer and Baum 2005):

$$P(\tau_{ij}) = \frac{1}{\overline{\tau_{ij}}} \int_0^{\tau_{ij}} P(\delta t_{ij}) d\delta t_{ij}. \tag{3.1}$$

Thus, depending on the properties of $P(\delta t_{ij})$, we could have a very large observation time τ_{ij} for the tie. As we will see in Chap. 4, the distribution for inter event times depends mostly on the average inter-event time $\overline{\delta t}_{ij}$, i.e. $P(\delta t_{ij}) = \mathcal{P}(\delta t_{ij}/\overline{\delta t}_{ij})$ where $\mathcal{P}(x)$ is a universal function (Miritello et al. 2011; Karsai et al. 2011). Thus, for a given $\overline{\delta t}_{ij}$, Eq. 3.1 can be written as:

$$P(\tau_{ij}|\overline{\delta t}_{ij}) = \frac{1}{\overline{\tau_{ij}}} \int_0^{\tau_{ij}} \mathcal{P}(\delta t_{ij}/\overline{\delta t}_{ij})\, d\delta t_{ij}. \tag{3.2}$$

If we suppose that each node chooses its tie activity from a distribution $\Pi(\overline{\delta t}_{ij})$ of average inter-event times across ties, then the probability to observe one of its ties at time τ is given by:

$$P(\tau_{ij}) = \int d\overline{\delta t}_{ij} \Pi(\overline{\delta t}_{ij}) P(\tau_{ij}|\overline{\delta t}_{ij}). \tag{3.3}$$

Thus, the growing function of the observed connectivity as a function of time is given by the cumulative distribution of $P(\tau)$:

$$k_i(t) = k_l(\infty) \int_0^t P(\tau) d\tau, \qquad (3.4)$$

where $k_i(\infty)$ is the total connectivity of node i in the observation time window. Since ties have very different $\overline{\delta t}_{ij}$ (or equivalently different weights $w_{ij} = T/\overline{\delta t}_{ij}$), which results in a heavy-tailed $\Pi(\overline{\delta t}_{ij})$, $P(\tau)$ is heavy tailed too. Thus aggregating over time the first appearance of the ties in the observation window will result in an apparent non-trivial time dependence $k_i(t)$, even if all the ties are active during the whole period. Equation 3.4 shows that one should be careful to consider the observed aggregate connectivity $k_i(t)$ as a proxy for social connectivity, since it is profoundly affected by the heterogeneous activity of human behavior.

Actually, the apparent growth $k(t) \sim t^{\gamma}$ can be observed also when the distribution of inter-event times is given by the exponential $P(\delta t|\overline{\delta t}) = e^{-\delta t/\overline{\delta t}}/\overline{\delta t}$ as well as the distribution for the average inter-event time $\Pi(\overline{\delta t}) = e^{-\overline{\delta t}/a}/a$. In this case from Eq. 3.4 we get:

$$k_i(t) = k_i(T) \left\{ 1 - 2\sqrt{\frac{t}{a}} K_1 \left(2\sqrt{\frac{t}{a}} \right) \right\}, \qquad 0 \leq t \leq T \qquad (3.5)$$

where $K_1(x)$ is the Modified Bessel Function of the second kind of order 1 (Abramowitz and Stegun 1972). As shown in Fig. 3.6a, also for this homogeneous (both in the events and in the ties properties) case, for a single user the number of observed ties grows in a non trivial way as a function of time, a behavior which extends further from $t = a$, where a is the average $\overline{\delta t}$.

The homogeneous case can be generalized to the case in which ties activity is much heterogeneous. This can be accomplished by considering a Gamma distribution, instead than an exponential, for the average inter-event times of $\overline{\delta t}$:

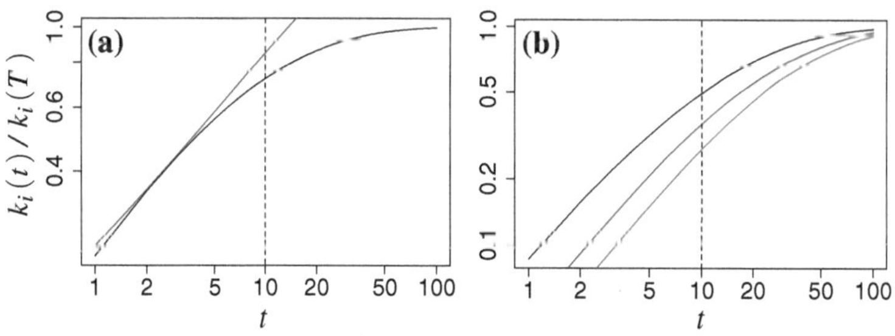

Fig. 3.6 Apparent growth of the observed connectivity given by Eq. (3.4) as a function of time for **a** an exponential distribution (*black curve*) and **b** a Gamma distribution of average inter-event times with $k = 2$ (*black curve*), $k = 3$ (*blue curve*) and $k = 4$ (*green curve*) and $\theta = a/k$. In both cases the average inter-event time of the distribution is $a = \overline{\delta t} = 10$ days and represented by the *vertical dashed line*. The *red line* in plot (**a**) represents a power-law fit for the initial growth of k given by $k_i(t) \sim t^{\gamma}$ with $\gamma = 0.53 \pm 0.02$

$$\Pi(\overline{\delta t}) = \frac{1}{\theta^k \Gamma(k)} x^{k-1} e^{-x/\theta}, \tag{3.6}$$

where $a = k\theta$ is the average $\overline{\delta t}$. From Eq. 3.4 we now get:

$$k_i(t) = k_i(T) \left\{ 1 - \frac{2}{\Gamma(k)} \left(\frac{t}{a}\right)^{k/2} K_k \left(2\sqrt{\frac{t}{a}} \right) \right\}, \tag{3.7}$$

where $K_k(x)$ is the Modified Bessel Function of the second kind of order k. Note that for $k = 1$ we recover the above case. As shown in Fig. 3.6b we obtain similar behavior as the previous case, provided the average $a = \overline{\delta t}$ is the same.
This result for a single user based on the universal bursty and heterogeneous activity in ties, together with the large heterogeneity found in social connectivity (which is related to $k_i(T)$) could explain the apparent non-trivial growth of the aggregate $k_i(t)$ observed in social networks (Krings et al. 2012) and highlights the importance of taking into consideration the heterogeneous human activity to define properly the way we measure and observe social networks.

3.2.2 Detection of Ties Creation/Removal

The effect of the bursty tie communication becomes critical in the study of allocation strategies in social networks, where the social connectivity plays a crucial role. In fact, to infer the existence of a tie thus detect social connectivity, it is not enough to observe a communication event within the observation window, but assess its formation and termination. The latter is a very challenging task and probably one of the main reasons why it is so difficult to model dynamic social networks. In fact, as we have seen in the previous section, due to the fact that the bursty behavior of human communication leads to a very slow revelation of ties over time, tie activity and tie creation/removal are often confused, especially when studying a network over short time periods. Therefore, to address this, we firstly need to consider a very long time windows. For this reason in the following we consider the same mobile phone network used in Sect. 3.1, but over a longer period of 19 months (instead than 11 months). Then, we need to develop a methodological approach to properly identify whether a tie has been actually formed or decayed, which is what we present is this section.

To separate tie activity from real formation/decay, we split our database into three time intervals of respectively 6, 7 and 6 months (Feb09–Jul09, Aug09–Feb10, Mar10–Aug10) and focus on the study of the tie formation/decay in the window Ω in the middle (see Fig. 3.7). The intervals before and after the observation window are used to asses whether each tie appears before and/or persist after Ω. Figure 3.7 shows the different situations that can occur for a given tie. We will consider that a tie $i \leftrightarrow j$ forms [case (b) and (c) in Fig. 3.7] or decays [case (a) and (c) in Fig. 3.7]

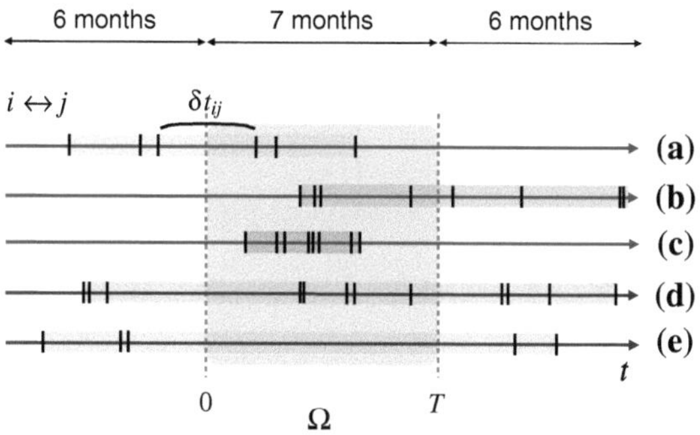

Fig. 3.7 Schematic view of the different situations of tie formation/decay and the interplay between tie communication patterns and tie formation/decay for a given observation time window Ω (*shadowed area*). Each *line* refers to a different tie while each *vertical segment* indicates a communication event between $i \leftrightarrow j$; δt_{ij} is the inter-event time in the $i \leftrightarrow j$ time series

within the observation window, if there is no recorded communication before or after respectively.

In our database, the 12.5 % of links belongs to the category (a), the 14.5 % to (b), the 22.2 % to (c) and the 47.3 % to (d). Of course, it is possible that even if there is no communication before/after the observation window, the tie is still active out of that period. This requires that ties would have an inter-event time δt_{ij} bigger than 7 months, i.e. case (e) in Fig. 3.7. However, in our database, only 3.5 % of the links have such long inter-event time, which assures the accuracy of our definition for tie decay/formation. To give significance to these findings, we only we keep *active* users, namely only those users who are involved (as calling or as called party) in, at least, one communication event in each of the three time intervals. Moreover, to avoid subscriptions/unsubscriptions performed just before/after the observation window Ω, we only consider users that appear at least one month before and are still active one month after Ω. This filtering is because the activity of nodes that subscribed/unsubscribed just before/after Ω may introduce spurious effects in the analysis of edge dynamics. Specifically, we would observe a rapid growth of their social network at the beginning of the observation window or a fast dissolution at its end (Gaito et al. 2012; Kikas et al. 2012). The latter filtering results in the removal of about the 17 % of nodes and the 37 % of reciprocated links. Panel (a) of Fig. 3.8 shows the ties formation and removal process within the observation window Ω for a single node in our data set. For each tie, a grey horizontal segments is drawn from the time of the first observed communication to the time of the last communication by considering also the time windows before and after Ω. According to our definition of tie formation/decay these segments represent the life-time of each social link in the observation window. The call activity is instead represented by the black dots, each one corresponding to one communication event. For all those ties that are not observed before and/or after Ω, the times at which the first and last communication are observed are represented by the red and blue vertical segments, respectively. As

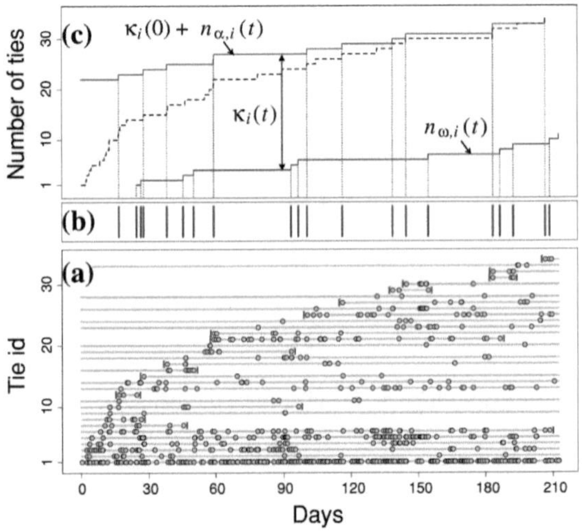

Fig. 3.8 Tie formation and removal process for a given individual: panel **a** shows the communication events of a given individual in our database with all his neighbors in the observation window. For each tie, a *circle* represents a call with the corresponding neighbor id. *Grey horizontal segments* are drawn from the first to the last observed communication event in each tie, considering also the time windows before and after the observation window. Panel **b** shows vertical segments for each tie formation (*red*) and tie decay (*blue*) events detected within the observation window. Using those events, panel **c** shows the aggregated number of open ties as a function of time (*red line*) and the aggregated number of closed ties (*blue line*). *Dashed line* is the apparent growth in the social connectivity obtained by the cumulative number of observed ties up to some time

one can see, we found a large difference between the number of ties *observed* up to time t (red segments until time t) and the number of ties already *formed* at time t (grey horizontal lines until time t). The same difference is observed between the number of ties observed until time t and the number of ties destroyed up to time t. This is, indeed, due to the entanglement between tie activity and tie dynamics discussed in the previous section. Due to the correlation between node activity and connectivity, this difference is larger for the social hubs. Note that for the node in Fig. 3.8 both curves coincide only from 5 months onwards and thus any shorter period of observation time will miss the real dynamics of tie formation/decay. This effect might explain why it has been so difficult to observe the edge dynamics in social networks. Due to the burstiness of tie communication, large inter-event times between interactions are likely and thus they might be unobserved or mistaken with tie decay or formation, especially if the observation window is short (order of months) (Akoglu and Dalvi 2010; Hidalgo and Rodriguez-Sickert 2008; Raeder et al. 2011).

3.3 Dynamical Communication Strategies

The procedure described above is the key of our analysis since it allows us to distinguish with high precision those connections that form or decay in the time window Ω from the one already formed before or persisting after. In fact, as mentioned in the Introduction, the high entanglement between tie dynamics and tie activity has been and still is one of the main impediments in the analysis of dynamical social networks. Separate these two time scales is crucial if we want to analyze the *instantaneous* social network, in contrast to the *aggregated* one, which is indeed the main goal of this thesis. In this section we will see that identifying the time instants of tie formation and removal is fundamental to analyze the way in which people dynamically allocate their attention across their contact network. In particular, since we are able to identify tie formation/removal events, we can calculate for each individual i what we call the *social capacity* $\kappa_i(t)$, which measures the number of active (or open) ties at any given instant t (see Fig. 3.8). It is important to stress that in principle $\kappa_i(t)$ is very different from the standard social connectivity $k_i(t)$ which measures the aggregated number of revealed ties up to time t (see Chap. 2) thus accounting also for those relationships which are not anymore active at time t. Social capacity and social connectivity are however related to each other. In fact, by aggregating the number of activated (deactivated) ties up to time t, denoted by $n_{\alpha,i}(t)$ ($n_{\omega,i}(t)$), we get that $k_i(T) = \kappa_i(0) + n_{\alpha,i}(T)$. Thus $k_i(T)$ is a combination of the social capacity κ_i and the *social activity* $n_{\alpha,i}$ in the observation period. Note that, since $\kappa_i(t)$ can vary along time, it might be independent of $n_{\alpha,i}(T)$ or $n_{\omega,i}(T)$, which allows a particular individual to follow different tie evolution strategies. Actually, as shown in Fig. 3.9a we observe a high heterogeneity in $n_{\alpha,i}$ and $n_{\omega,i}$: while on average people create/destroy about 8 (reciprocated) ties in a period of 7 months, the 20 % of users in our database add or remove more than 15 ties in that period. This is a relatively large number for a mobile-phone communication, where much more effort is required to establish and maintain a tie if compared to online communication networks as Twitter or Facebook, which are often used to collect as many friends and followers as possible. On average, we find that the number of added (removed) ties $n_{\alpha,i}$ ($n_{\omega,i}$) almost equals $k_i/2$, indicating that a large fraction of the standard measure of social connectivity k_i is given by newly formed or removed connections. This suggests that the aggregated social connectivity k_i usually overestimates the real human social capacity of maintaining social relationships. The imbalance between the activated ties $n_{\alpha,i}$ and the deactivated ones $n_{\omega,i}$ allows us to measure how the social capacity changes in time. At the end of the observation period, the change in the social capacity is given by $\kappa_i(T) = \kappa_i(0) + n_{\alpha,i}(T) - n_{i,\omega}(T)$. Interestingly, we find that the large majority of users in our database show $n_{\alpha,i}(T) \simeq n_{\omega,i}(T)$ (Fig. 3.9c). As a simple explanatory hypothesis for the observed $n_{\alpha,i} \simeq n_{\omega,i}$ behavior, one might suggest a sort of conservation principle in social attention, where the number of outgoing ties equals the number of incoming ties, such that the total number of open relationships in a given time window Ω remains almost constant. This conservation of social capacity not only happens at this particular time scale T but also instantaneously: as shown in

Fig. 3.9 we find that for 80 % of the users tie formation/destruction happens linearly in time so that $n_{i,\alpha}(t) = \alpha_i t$ and $n_{i,\omega} = \omega_i t$, where the rate α_i at which ties are formed equals the one at which ties are removed ω_i. To compute the rates α_i and ω_i, for each user we model the time sequences of tie arrivals and removals as linear processes, thus apply to them a linear regression fit. In order for the regression to be significant, we only keep those nodes that form and remove at least 5 connections during the whole period (e.g. $n_{\alpha,i} \geq 5$ and $n_{\omega,i} \geq 5$), which in our data set corresponds to almost the 20 % of nodes. As shown in Fig. 3.9b, for these users we obtain a $p - value \leq 0.05$ for the 99.2 % of the cases (further details about the analysis performed will be provided in the next section). The fact that $\alpha_i \simeq \omega_i$ has a remarkable consequence since it means that people add connections at a constant rate α_i and, at the same time, they remove connections at a similar rate ω_i. This implies that the individual social capacity remains almost constant throughout the time period since $\kappa_i(t) = \kappa_i(0) + \alpha_i t - \omega_i t \simeq \kappa_i(0)$ signaling that people tend to balance the formation/removal of social connections such that the instantaneous number of their open connections remains almost stable over time (Miritello et al. 2012).

Note that our finding explains many observations in the literature that the distribution of connectivity in social networks seems to be stable in time but the neighbors of a given node change a lot from one time window to the other (Hidalgo and Rodriguez-Sickert 2008; Kossinets and Watts 2006). In fact, we find that the average social persistence p_i, i.e. the fraction of neighbors that remain at the end of the 7 months observation time window is around 75 %, meaning that users renew their social circle slowly, in line with studies in off-line social networks (Burt 2000). This value is much larger than what is expected in a random model where every tie has the same probability to be deactivated. To check this we simulated a process in which each individual (characterized by its actual k_i, κ_i and n_α) is allowed to create and remove connections according to its real tie activation/deactivation sequence of events, but where added and removed ties are now randomly chosen among the whole set of neighbors. This yields to an average social persistence $\overline{p}_i = 50\,\%$, which is significantly lower than the real one. This result indicates that the way in which people activate and deactivate ties from their social network is far from random; instead, some existing ties are more probable to be deactivated than others. In Sect. 3.5 we will return to this result and show one of the possible mechanisms that could give rise to the observed bias.

3.3.1 Statistical Evidence for the Conservation of Social Capacity

In the following, we describe the procedure that we used to calculate the two rates α_i and ω_i for all the users in our database and the analysis performed to state that individuals create and destroy ties at the same rate. We will also discuss the null model used to assess the statistical significance to our findings. For most users in our database, the number of added and removed connections ($n_{i,\alpha}$ and $n_{i,\omega}$) is very

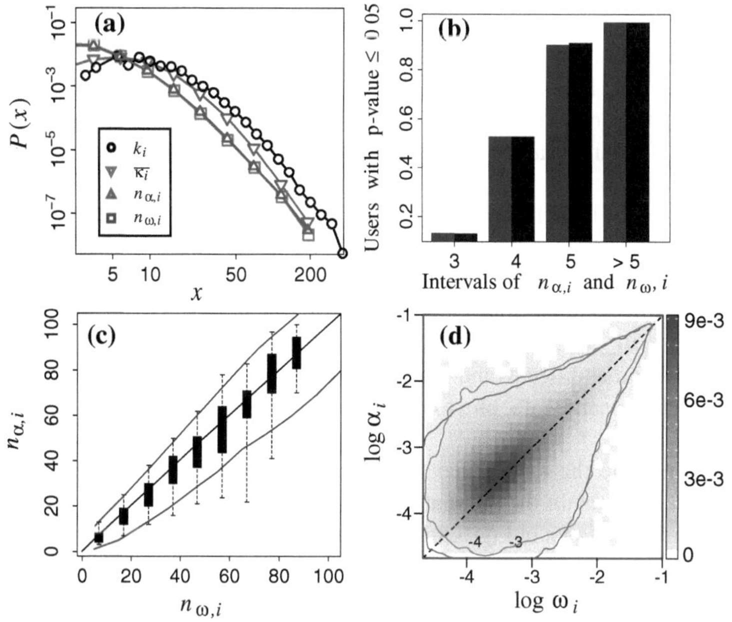

Fig. 3.9 **a** Distribution of the social connectivity k_i (*black circles*), social capacity κ_i (*green downside triangles*) number of created ties $n_{\alpha,i}$ (*red upside triangles*) and number of deleted ties $n_{\omega,i}$ (*blue squares*). We observe that, on average, both $n_{\alpha,i}$ and $n_{\omega,i}$ almost equal $k_i/2$, which suggests that a large fraction of the aggregated social connectivity is due to newly formed or removed ties. **b** P-values of the linear fit used to compute the rates of tie creation α_i and removal ω_i for groups of users with a different number of created and deleted connections (respectively $n_{\alpha,i}$ and $n_{\omega,i}$). We observe that for the 99.2 % of users with both $n_{\alpha,i}$ and $n_{\omega,i} \geq 5$ the obtained p-value is always smaller than 0.05. **c** Positive correlation between the number of created $n_{\alpha,i}$ and removed $n_{\omega,i}$ connections with a linear correlation coefficient of 0.87. The results from the PCA analysis indicate that the 93 % of the variation can be explained by the first component with a standard deviation of 1.81 in the (0.70, 0.71) direction. This result is shown in the box plot, where the *bottom* and the *top* of the boxes correspond to the 25*th* and the 75*th* quantiles respectively, while the band near the middle is the 50*th* percentile (median) of the distribution. The *down* and *top* of the whiskers represent the 5*th* and 95*th* percentiles. The line y = x lies between the 9th and the 91st percentiles, thus capturing the 90 % of the distribution in correspondence of each box. The blue solid lines refer to the 5th and 95th percentiles of random generated n_α and n_ω from a Poisson distribution with n_α taken as the expected number of events in a given time interval of length T. **d** Log-density plot $\rho(n_{\alpha,i}, n_{\omega,i})$ for users with more than 5 formed and removed ties. The *dashed lines* refers to $\alpha_i = \omega_i$ while the *solid curves* correspond to the contour lines $\rho = 0.03$ for the density of actual values of the rates (*red*) and the ones obtained in the Poissonian model (*blue*)

small and then we get large differences between the values of α_i and ω_i. We tested that our results are comparable statistically to a null model in which ties are formed and destroyed in the observation window Ω according to two realization of a Poisson process with the same rate $\alpha = \omega$. The choice of Poisson process as the renewal process that describes the formation and decay process is supported by the bounded probability distribution for the inter-event times between formation/decay of events

seen in Sect. 3.2.1. This is of course an approximation, because the probability of bursts of formation/decay events is larger than what predicted by the exponential distribution of the Poisson process. The approximation works better for large times or number of events, since in that limit the strong decay of the inter-event time distribution for large values makes the process to converge to the behavior of a Poisson process very quickly by means of the Central Limit Theorem (Cox 1970).
To incorporate the large heterogeneity of social activity observed in our database (see Fig. 3.9a), we take as input for our null model the actual values of $n_{i,\omega}$. Analogously, we have also done simulations by considering $n_{i,\alpha}$ and obtained the same results.
Our Monte Carlo simulations can be summarized as follows: for every user i we consider $\lambda_i = n_{i,\omega}/212$ (where 212 is the number of days within Ω) as the daily rate for tie formation and decay and simulate two Poisson processes in the observation window with the same rate, one for the formation of ties and the other for the decay of ties.
We then calculate the aggregated number of events $\hat{n}_{i,\alpha}$ and $\hat{n}_{i,\omega}$ and fit them to linear models to obtain the simulated $\hat{\alpha}_i$ and $\hat{\omega}_i$ for those cases in which $\hat{n}_{i,\alpha} \geq 5$ and $\hat{n}_{i,\omega} \geq 5$. As shown in Fig. 3.9c the empirical values of $n_{i,\alpha}$ observed in our data can be well explained by the simulations, suggesting that our model works well at the particular scale considered. In addition, we also find a good agreement between the measured values of α_i and ω_i and the simulations (Fig. 3.9d), despite of the small amount of outliers that cannot be explained by our model.

3.3.2 Dynamics of Tie Creation/Removal

We have seen that the growth of the number of added and removed connections can be described by a linear process having rates α_i and ω_i, respectively. However, as panel

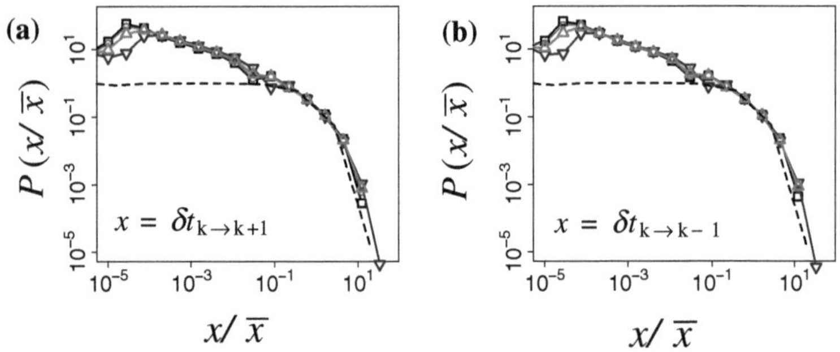

Fig. 3.10 (Rescaled) Distribution of the time gap between edge creation (**a**) and edge removal (**b**). Each *curve* refers to a group of nodes with a different activity rate α_i, where groups have been obtained according to the quartiles of α_i for the whole population. The *dashed line* in both panels correspond to an exponential distribution with the same mean

(b) of Fig. 3.8 suggests, it is important to note that the distribution of the time gap between creation and removing of ties is not an homogeneous Poissonian process. To characterize the temporal evolution of ties we analyze the distribution of the inter-event times elapsed between consecutive additions or deletions. We denote these time differences respectively $\delta t_{k,k+1}$ and $\delta t_{k,k-1}$ since they increase and decrease the social connectivity k of one unit. Our results in Fig. 3.10 indicate a heterogenous pattern of activity and, despite the exponential cut-off, small inter-event times are significantly more probable than an exponential distribution. This is in line with previous research that has shown that the nature of link creation process in online social networks is broad and follows a power-law distribution (Gaito et al. 2012; Kikas et al. 2012; Leskovec et al. 2008). In these studies the bursty behavior of tie creation is usually associated to an *acceleration* (or exploration) phase in which tie formation rapidly increases. Since the trains of bursts are more likely to appear right after the users joined the network, this acceleration phase is usually related to the registration time of the user. However, here we show that the observed distribution remains robust with respect to this effect since, as mentioned above, we are considering only users who are active enough time before and persist enough time after the observation window Ω. More interestingly, we show that tie creation dynamics does not depend on activity rate α_i of tie creation. This is shown in Fig. 3.10, where we aggregate the whole population of users in different groups depending on the rate α_i of their activity (where the groups have been defined according to the quartiles of the distribution of α_i) and analyze the distribution of $\delta t_{k,k+1}$ and $\delta t_{k,k-1}$ for each group. As one can see, both distributions are robust with respect to α_i, indicating that on average the way in which individuals create social relationships does not depend on the velocity at which such connections are added. For the equivalence $\alpha_i \simeq \omega_i$ seen in the previous section, it nether depends on the rate of tie removal. In this sense, this universal behavior is reminiscent of the universal behavior of bursty activity within the ties.

3.3.3 Social Capacity and Social Activity

According to our results presented in Sect. 3.3, each user can be characterized in terms of his *social capacity* κ_i and *social activity* $n_{\alpha,i}$ (or equivalently α_i). These two quantities give us information about two related although different features of an individual. While the social capacity is a measure of the number of relations that a user maintains in time, the social activity instead characterizes the number of relations a user establishes and at what rate. However, by applying Principal Component Analysis (PCA) analysis, we observe that for most of the individuals $n_{\alpha,i} \simeq \beta \overline{\kappa_i}$ with $\beta = 0.75$ meaning that the number of created connections tends to be proportional to the social capacity, as shown in Fig. 3.11. The correlation between κ_i and $n_{\alpha,i}$ resembles the *preferential attachment* process discussed in Chap. 2, by which tie formation is more probable for more connected individuals (Barabási and Albert 1999). Note however that we find that tie formation is proportional to a conserved quantity and thus grows linearly in time. Moreover, there is a corresponding *prefer-*

ential de-attachment meaning that individuals with large $\overline{\kappa_i}$ are also more likely to destroy ties.
Although the dependence $n_{\alpha,i} \simeq \beta \overline{\kappa_i}$ explains most of the observed behavior (80 % of variance in PCA), there is however a large variability in our database so that tie evolution cannot be explained solely by $\overline{\kappa_i}$. We encode the disparity between the social capacity and social activity in the ratio $\gamma_i = n_{\alpha,i}/\overline{\kappa_i}$, which we dub as social strategy of individual communication. The parameter γ gives information about the equilibrium between the social capacity and the social activity for a given node: for $\gamma_i \simeq \beta$ (the average behavior), users have a *normal* or *balanced* social strategy between their social capacity and activity. Outside this group we find those users with $\gamma_i \ll \beta$ that activate/deactivate a small number of connections compared to their social capacity, or users with $\gamma_i \gg \beta$ who have a large social activity compared to their social capacity. We refer to these two strategies as *social keeping* ($\gamma_i \ll \beta$), meaning that these individuals keep a very stable social circle, and *social wandering* ($\gamma_i \gg \beta$), meaning that these individuals activate new ties elations and deactivate existing ones at a high pace. It is very important to note that, despite the strict relation between γ_i (or both $n_{\alpha,i}$ and κ_i) and the social connectivity k_i, individual social strategies of communication can not be identified only by means of k_i.

This becomes clear by looking at panel C of Fig. 3.11, where we show that users with exactly the same k_i (dashed curves) can be characterized by very different combinations of social capacity and social activity. As we will see in the following section, the implications of this result go beyond the characterization of how people allocate time and resources across their social circle. In fact, the adoption of one strategy or the other is strictly related to the topology of the local network surrounding the individual, with important implications in terms of information diffusion.

3.4 Social Strategies and Network Topology

After characterizing the social strategy of each user in our database, we investigate whether the nature of such strategy depends on the properties of the individual local network. Specifically, for each individual we measure the *persistence* of his neighborhood and *(ii)* his *clustering coefficient* or transitivity. The *persistence* p_i gathers information about the continuity of the social relationships of an individual and, consequently, about the speed at which he renews his social circle. It is measured as $p_i(T) = |E_i(T_0) \cap E_i(T)| \,/\, |E_i(T_0)|$, where $E_i(T)$ ($E_i(T_0)$) is the set of ties that user i has a time T (T_0) and can take values between 0 (all the contacts has been changed after a period T) and 1 (all the contacts persist after T). The clustering coefficient c_i, as discussed in Chap. 2, measures how likely individual's neighbors are to be connected to each other. Our results are shown in Fig. 3.11, where we depict the snapshots of the neighborhood of two different individuals at different time instants. Note that, as shown in panel C, although the chosen individuals have the same social connectivity k_i, they show two different social strategies being one (user A) a *social wanderer* and the other (user B) a *social keeper*. These snapshots clearly show the

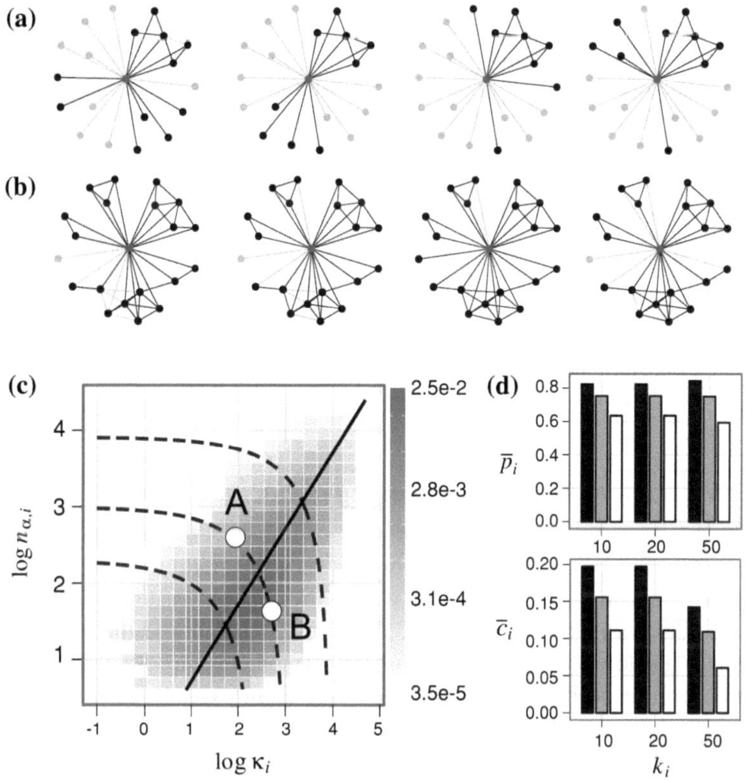

Fig. 3.11 Heterogeneity of dynamical social strategies: **a** and **b** show different snapshots of the neighborhood of two different individuals (in *red*) at four equally spaced times in the observation time window $t = 52, 105, 158$, and 211 days. Each *black* (*grey*) *line* corresponds to an *open* (*closed*) tie at that particular instant. **c** Log-density plot of the social activity $n_{\alpha,i}$ as a function of the social capacity κ_i for each individual in our database. *Solid line* corresponds to the line $n_{\alpha,i} = 0.75\overline{\kappa}_i$ obtained through PCA. *Dashed curves* are the iso-connectivity lines $k_i = \overline{\kappa_i} + n_{\alpha,i}$ for $k_i = 10, 20, 50$. **d** shows the average value for the persistence p_i and clustering coefficient c_i for three groups of equal connectivity (*dashed lines* in panel **c**) but for different quantiles of γ_i. Specifically, $\gamma_i < 0.43$ (*black*), $0.43 < \gamma_i < 0.88$ (*gray*) and $\gamma_i > 0.88$ (*white*)

difference between the local network of the social wanderer and the one of the social keeper: while the former network appears quite volatile and sparse, the latter looks very stable in time and highly connected. In particular, as shown in panel D, for all users we find that although on average they show a 75 % persistence in their social circle (in 7 months), this value rises up to 80 % for social keepers and is only 60 % for social wanderers. A similar dependency is found for the clustering coefficient c_i: for a fixed k_i the clustering coefficient for social keepers doubles that of social wanderers, meaning that for equal k_i the former have less distinct social contexts or structural diversity than the latter users, which is in line with previous results (Ugander et al. 2012).

Taken together, these findings document an important difference between possible social dynamics. On one hand, social wanderers are mostly navigating the network looking for new social connections and/or information, resulting in a larger structural diversity. On the other hand social keepers are more conservative people, who pay mainly attention to a large proportion of their stable social network, which also turns out to be tightly linked together: the friends of their friends are also friends to each other. Thus for a given k_i social wanderers trade off novelty for attention resulting in a high volatile social strategy, while social keepers display a more stable and less diverse social structure around them. These results constitute a step forward into the understanding of how to model dynamical social networks, since they allow us to go beyond the traditional characterization of individuals static aggregated variables like k_i and the static topological structure of networks.

3.4.1 Assortative Mixing in Dynamical Social Strategies

In Chap. 2 we have seen that social networks show assortative mixing in the social connectivity, meaning that individuals with many connections tend to be connected to other nodes with many connections. We have also mentioned that the assortative nature of social networks emerges as a natural phenomenon in both offline and online networks with respect to a variety of attributes, from psychological states such as loneliness or happiness (Bliss et al. 2012; McPherson et al. 2001) or health attributes and habits (Christakis and Fowler 2007, 2008) to sociodemographic features such age or race (Ibarra 2002; Mollica et al. 2003). In this section, we show that this tendency also applies in the context of dynamical social communication.

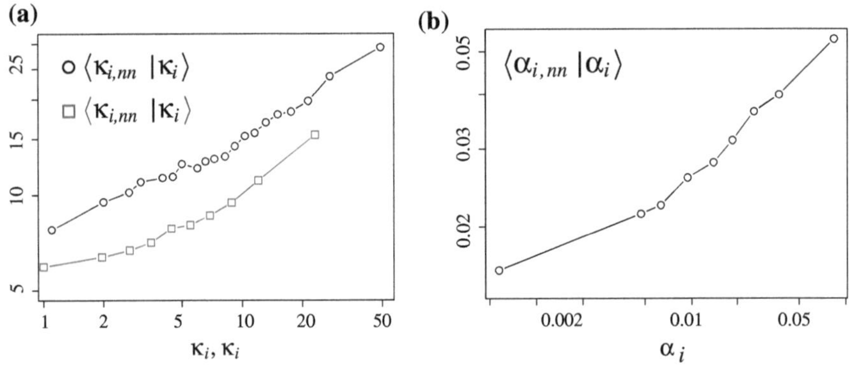

Fig. 3.12 Average social connectivity k_i, social capacity κ_i (**a**) and social activity rate α_i (**b**) for users (x axis) and their neighborhood (y axis). The results show that there is assortative mixing in the social connectivity k_i, social capacity κ and social activity α. Users with a high capacity to maintain and establish social ties are more like to connect with people with the same capabilities

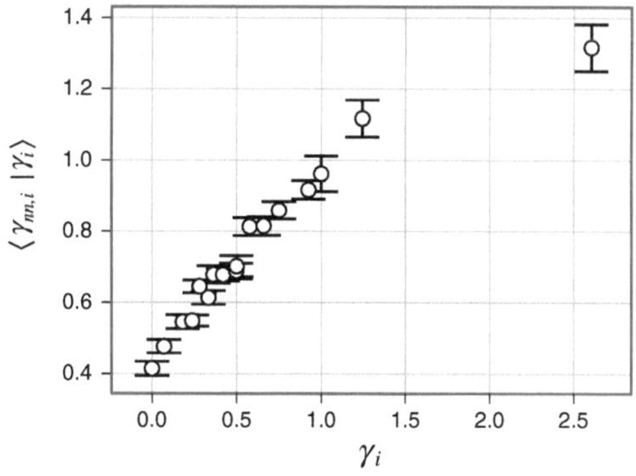

Fig. 3.13 Assortativity of the dynamical social strategies. Average value of the parameter γ_i for the neighbors of an individual as a function of her own value of γ_i. A clear growth can be seen, indicating a strong assortativity of the social dynamical strategy in our database

First, as shown in Fig. 3.12a, we observe assortative mixing not only in the social connectivity k_i, which is in line with previous studies (Newman 2002a), but also in the *social capacity* κ_i, meaning that people maintaining a large (small) number of open relationships tend to be connected to other people that also maintain a large (small) social circle. Analogously, people with a high capacity to establish and remove many connections are more likely to connect with people with the same characteristics, which reflects in the observed assortative mixing of the *social activity* (Fig. 3.12b). More interestingly, the assortative character of dynamical social strategies is observed also in the parameter γ_i (see Fig. 3.13), indicating that social wanderers (as well as social keepers) tend to gather together. This suggests that the large volatility observed in the neighborhood of social wanderers also extends to large proportions of the network around them and the same applies for social keepers.

In this picture, the global network consists of almost static zones of social keepers and high volatile clusters of social wanderers. A possible explanation for the observed behavior might be the following. On one hand, since social keepers invest the larger amount of their energy/cognitive ability in keeping stable social contacts, they have lower reservoir or interest for exploratory relationships. This strategy of communication would lead them to reject or avoid most of the connections that social explorers, instead, might try to establish. On the other hand, if these latter connections are unlikely to happen, the social circle of social explorers ends up being restricted mostly to other social explorers. Note that, although the adoption of social strategies does not depend on the social connectivity (as seen in the previous section), their assortative nature reminds the one observed for this latter quantity in a wide range of real social networks, which somehow characterize the static individual strategy of communication. The reason why these strategic has not surfaced yet stems from the fact that, as discussed in Sect. 2.3, attention and social capacity have been considered infinite, and therefore strategic considerations in the acquisition of a new link were out of the discussion. Once this limited resource is identified, it follows that individuals would establish different strategies to overcome it. If we were to follow Grannoveter's (Granovetter 1973) and Burt's advice (Burt 1992), thus go about gathering weak links, and occupying structural wholes, to what extent do we follow it? Without a limited resource in social capacity these questions make little sense (Hardin 1968).

3.4.1.1 Social Strategies in Facebook

As already mentioned, most of the results presented in this thesis, as well as the observed communication strategies discussed above, have been obtained by analyzing a mobile phone dataset. In this section we show that such strategies also arise in online settings, although a more complete study of online social networks is beyond our scope. By address this, we use the Facebook dataset described in Appendix A, where a communication event between two users represents an interactions through the Facebook wall. Specifically, we consider only communication events between those Facebook users in our dataset that show any activity in the 7-months observation window Ω (see Fig. A.3 in Appendix A) and, in a way analogous to what done for the mobile phone network, we only retain users that are active 20 days before and 20 days after Ω.

We find that, on average, users interact with $\langle k_i(T)\rangle = 6.15$ users in the period Ω. The average social capacity is instead $\langle \kappa_i(t)\rangle = 3.23$ while the average number of formed and decayed ties is, respectively, $\langle n_{\alpha,i}(T)\rangle = 3.01$ and $\langle n_{\omega,i}(T)\rangle = 3.02$. Nevertheless, Facebook users show a lower level of activity: for example, 40 % of the users are involved in less than 10 communication events through the wall in seven months (while in the mobile phone data the average number of calls exchanged per user was ~700 in 7 months). Thus, to determine the social dynamical strategies in Facebook data we concentrate on those users that show a moderate level of communication, i.e. those that have more than 10 events in the 7 months of Ω. As shown in Fig. 3.14, we observe that for those users $n_{\alpha,i}(T) \simeq n_{\omega,i}(T)$ and $\alpha_i \simeq \omega_i$. Specifically, analogously to the analysis done for the mobile data set, we find that by applying PCA the relationship between social capacity and social activity is given by $n_{\alpha,i} = 1.04\overline{\kappa}_i$, which explains the 81 % of the variance. These results indicate that Facebook users, as well as mobile phone ones, tend to create and destroy social connections linearly in time at a (almost) equal rate, which leads to a conservation of the open connections $\kappa_i(t)$ in time. In addition, analogously to the mobile phone case, we observe a large assortativity of the social dynamical strategies, i.e. individuals with low γ (social keepers) tend to gather in the social network, while social wanderers tend to communicate between them (see Fig. 3.15).

Note that, as mentioned above, the average social capacity is roughly half of the social connectivity in 7 months, suggesting that also in this case the standard aggregated social connectivity k_i overestimates the actual individual capacity to maintain social relationships. On one hand, this finding confirms previous studies showing that, although people on Facebook have on average hundreds of friends, they are only in direct contact (direct exchange of information such as wall conversation) with a small proportion of them which lies around 10–30 (Marlow 2009). On the other side, however, our analysis shows that this number decreases even further when, after disentangling the tie activity from their formation/decay dynamics, the instantaneous, instead than the aggregated, network is considered.

3.4.2 Social Strategies and Information Diffusion

The existence of social strategies and, more importantly, the fact that they are assortative, motivated us to investigate what implications this has in terms of information diffusion. Indeed, a relation between social strategies and accessibility to information could have important implications on the current modeling of many real processes such as diffusion of innovations (Rogers 1995), news and opinions (Adar et al. 2004; Leskovec et al. 2009) because it would tell us which users have higher capacity to get or spread information, based on the way in which they communicate. For example, since some regions of the network are volatile whereas others are static, one may wonder whether this affects or not the propagation of a piece of information within the network.

To this end, we simulate the SI model (Murray 1993) in the mobile phone network under investigation, by considering the real-time sequence of communication events (Miritello et al. 2011; Karsai et al. 2011; Zhao and Oliver 2010). According to this process, a population of N individuals is divided into two states: susceptible (S) and infected (I). Susceptible individuals can become infected with a given transmission rate λ (usually equal to 1) any time they interact with an infected individual. The SI model is the simplest formulation of a class of epidemic models known as compartmental models, which are usually used to simulate the spread of epidemic

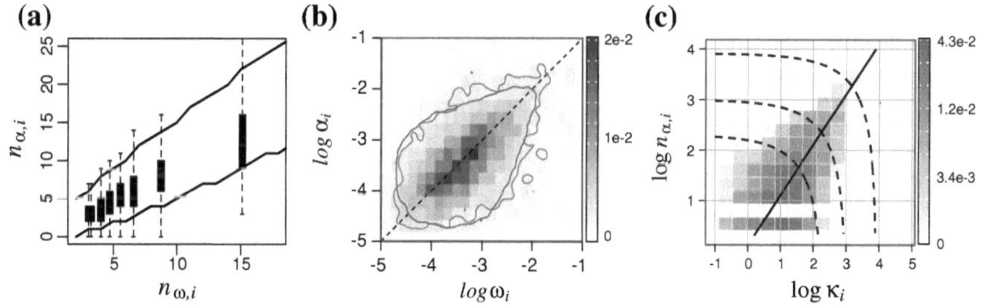

Fig. 3.14 Social dynamics in the Facebook database: **a** Relationship between the number of formed $n_{\alpha,i}$ and decayed $n_{\omega,i}$ ties in the observation window. The *box* plot shows the 25 and 75 % percentiles (*filled box*) and 5 and 95 % percentiles (*whiskers*), the *solid black line* is the relationship $n_{\alpha,i} = n_{\omega,i}$ and the *blue curves* correspond to the 5 and 95 % percentiles of the corresponding Poisson model described in Sect. 3.3.1 for the mobile dataset. **b** Density plot $\rho(n_{\omega,i}, n_{\alpha,i})$ for the users with more than 2 ties formed and decayed. The *dashed line* represents $\alpha_i = \omega_i$ while the *curves* correspond to the contour lines $\rho = 0.03$ for the density of actual values of the rates (*red*) and the ones obtained in the Poissonian model in Sect. 3.3.1 (*blue*). **c** Log-density plot of the social activity $n_{\alpha,i}$ and the social capacity $\overline{\kappa}_i$. The *dashed lines* correspond to the iso-connectivity lines $k_i(T) = 10, 20, 50$ and the *solid line* is the relationship $n_{\alpha,i} = 1.04\overline{\kappa}_i$ obtained through PCA that explains 81 % of the variance

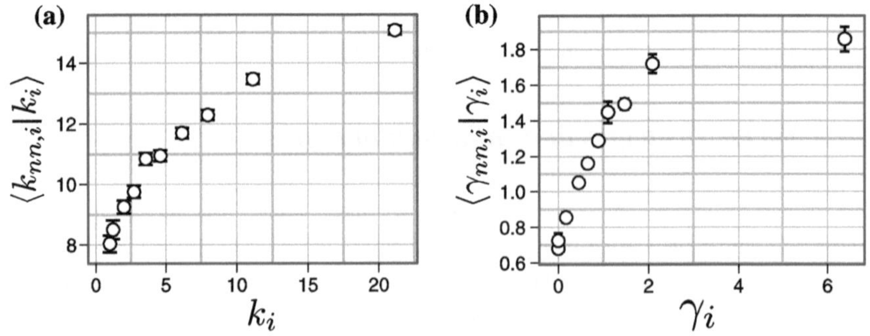

Fig. 3.15 **a** Average next neighbor connectivity of a node $k_{nn,i}$ as a function of its own connectivity k_i for the 10^4 users in the Facebook dataset. A clear grown can be seen, signaling an assortativity in this social network. **b** Average value of the parameter γ_i for the neighbors of an individual as a function of his own value of γ_i. Similarly to k_i we observe a growth which indicates a strong assortativity of the social dynamical strategies

diseases (Anderson and May 1992). Compartmental models, however, have been largely used also to model information diffusion processes, by means of the analogy between the spreading of an infectious disease and the dissemination of information (Daley and Kendall 1964; Goffman and Newill 1964). A detailed analysis of epidemic models and how they have been used to simulate information diffusion on social networks will be provided in Chap. 5. For the purpose of this section, what is important is that according to this process all the nodes will be eventually infected, with a speed that only depends on the network structure and the epidemic dies once the whole population has been reached. Moreover, simulating the SI model on the real series on the real-time sequences of communication events in our CDR, allows us to take into account also all those temporal inhomogeneities such as correlations or burstiness (see Chap. 2), in a way analogous to the one described in previous works (Karsai et al. 2011; Miritello et al. 2011; Vázquez et al. 2007). We start the

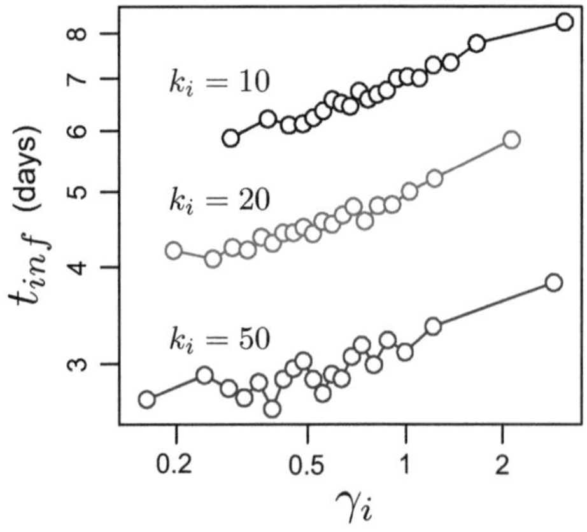

Fig. 3.16 Average difference of infection time (in days) as a function of the social strategy γ_i for each of the iso-connectivity groups in Fig. 3.11. The relative difference calculated with respect to the average infection time for each of the groups. Social wanderers (large γ_i) exhibit longer time to receive the information than social keepers (small γ_i)

model by infecting a random node (or seed) at a random instant and considering all other nodes as susceptible. In each call, an infected node infect a susceptible one (or propagate a piece of information) with probability $\lambda = 1$. Due to the synchronous nature of the phone communication, we assume that this happens regardless of who initiates the call. We then repeat the simulation for 10^4 randomly chosen seeds. For each realization of the process, we measure the time at which a particular individual receives the information from the starting time of the infection process. The result is shown in Fig. 3.16 where we plot the average relative time difference Δt_{inf} for users grouped by social connectivity, as a function of the average γ_i. The relative difference has been calculated with respect to the average infection time of each of the groups. Interestingly, we observe that when we control for k_i, social wanderers exhibit a relatively longer time to receive information spreading in the network compared to social keepers.

This result can be related to what observed at the tie level by Onnela et al. (Onnela et al. 2007) where an individual social strategy is assigned by aggregating the total communication time. The authors showed that in terms of information diffusion, weak ties are ineffective at information transfer because of the small amount of communication time. Here we see that this effect is amplified in social wanderers as (i) most of their ties are weak, and (ii) they are connected to other social wanderer with weak tie access to the rest of the world. Whereas in Onnela et al.'s contribution, both weak ties and strong ties are in similar level of performance regarding information transmission, the assortative nature of the communication strategies breaks the tie in favor of the strong ties. Moreover, contrary to the common belief that exploratory strategies would provide advantages in individual information access, the fact that they cluster together exacerbate the volatility of the interactions and produce relative inefficiencies in access to information (Miritello et al. 2012).

3.4.3 Lifetime Evolution and Sex Differences

Previous studies have shown behavioral and social differences in the use of mobile phones depending on the demographics (gender, age, socioeconomic status) of the individuals (Blumenstock and Eagle 2010; Frías-Martínez et al. 2010; Palchykov et al. 2012). This type of cross-sectional analysis has the potential to provide insight into the underlying dynamics of human behavior and in the understanding of status-based differences between individuals or countries, e.g. differences in the usage of technology or in the dynamics of sharing.

Here we study whether the dynamical social strategy of communication depend on the age and the gender of individuals. Interestingly, we found that although social capacity and strategy remain mostly stable over the observation time window, they tend to change on the longer time scale, i.e. during the individual life course. In Fig. 3.17a we show the average social connectivity k_i as a function of the average value of γ_i for groups of individuals with different age.

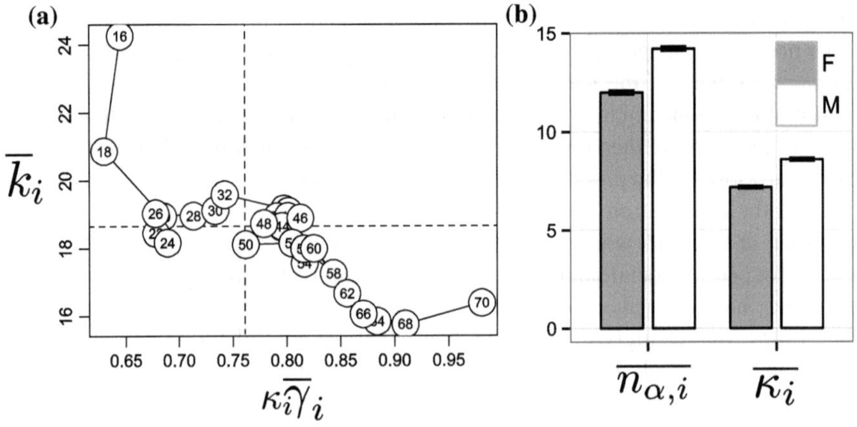

Fig. 3.17 **a** Average value of the social connectivity $\overline{k}_i$ and the social dynamics $\overline{\gamma}_i = n_{\alpha,i}/\kappa_i$ for group of users with different age, reported within the *circles*. *Dashed lines* correspond to the average of κ_i and $n_{\alpha,i}$ in the complete database. **b** Average values for the activity and capacity of users grouped by gender, where the M states for males and the F for females

Reminding that $k_i = \kappa_i + n_{\alpha,i}$, this result shows a decrease in both the social capacity and activity, is in line with previous studies (Kahn and Antonucci 1980; Seeman et al. 2012; Tilburg 1998) that also offer a possible explanation for such a behavior. Specifically, changes in egocentric network size across the individual lifespan are usually associated to both experiencing age-specific life events and social goals (Wrzus et al. 2012). On one hand, both normative events such as marriage, parenthood or retirement and non-normative events such as contracting a rare disease or winning the lottery, may cause a decrease or increase in the number of active ties (Morgan and March 1992). On the other hand, according to socioemotional selectivity theory (Carstensen et al. 1999), personal network size may be related also to goal hierarchies over the entire life which change when boundaries on time are perceived (Suitor et al. 1997). In particular, during adolescence and young adulthood a large social circle reflects the fact that information acquisition goals are more predominant than in other periods of life. After young adulthood, however, when people perceive remaining lifetime as limited, emotion regulation goals become much more important and people concentrate their attention in close relationships (Carstensen 1991). Other studies relate the decrease in the social engagement (number of social contacts, interaction activity, frequency of communication) across the individual lifespan, to a decrease in the cognitive capacity (Seeman et al. 2012).

However, our interpretation of k_i as a combination of n_α and κ_i allows us to better understand the change in social network size across the individual lifespan and its relation with individual communication strategies. By looking at Fig. 3.17a it seems that different behaviors emerge, in correspondence to different stages in people life: adolescence (16–19), young adulthood (20–32), adulthood (33–56) and old age (>56). In particular, young people and adults show a quite stable behavior which leads to reasonably defined clusters with large social connectivity and a large capacity to maintain open relationships; however, both properties decrease in the adulthood. A possible explanation for the observed behavior is that while young people tend to use mobile phone to pay attention to their social network and maintain stable connections, this behavior is less and less important for older people who conserve a smaller stable social structure compared to their social activity, maybe due to the appearance of other uses (professional contacts) or priorities (access to information). The behavior strongly differs from what observed before and after the adulthood periods. While adolescence is characterized by large values of connectivity and high capacity to maintain open relationships, during old age an ongoing decrease of k_i is observed together with a fast increase of γ_i. This latter suggests that as people get older, the number of social relationships they can maintain over time become increasingly smaller than the total number of relationships they keep contact with. Figure 3.17b shows the average values of both the social capacity and the social activity for users grouped by gender. In line with previous studies using mobile phone records (Frías-Martínez et al. 2010), we found that on average women have less capacity to maintain big social circles than men. Interestingly, we also observe a smaller social activity for women than for men indicating that, on average, women also display a smaller capacity (or interest) in establishing or removing social connections from their network (Miritello et al. 2012).

3.5 Dynamical Granovetter Effect

In Sect. 3.4 we have seen that, although users establish and remove many connections in time (on average half of their social connectivity), after a period of 7 months they maintain on average the 75 % of their initial social circle. This suggests that there is an inherent dynamics of tie creation and removal which makes the process of social circle renewal far from random. In Sect. 3.3 we have also checked the latter statement by simulating a process in which users are allowed to add and remove connections randomly, which led to a value for the average persistence of about the 50 % which is significantly lower than the one observed for the real data.

The fact that some connections are much more likely to be added or removed than others is not new in the field of social network (Adamic and Adar 2003; Getoor and Diehl 2005; Hidalgo and Rodriguez-Sickert 2008; Kossinets and Watts 2006; Liben-Nowell and Kleinberg 2007; Raeder et al. 2011; Roth et al. 2010) and a deeper analysis of this issue will be also the main topic of the next chapter. Nevertheless, here we want to address a related issue and try to understand what criterium people follow in the process of tie creation/removal. Among other mechanisms, individuals may for example decide to establish ties outside their current social circle (Burt 1992; Granovetter 1973) or they may choose new acquaintances who are friends of friends, a process known as *triadic closure* (Coleman 1988a; Rapoport 1953).

Understanding the very dynamics which regulates the process of why an individual decides to add or remove a social tie is a very complex process which has been of interest of many studies (Burt 2001b; Granovetter 1973; Kossinets and Watts 2006; Podonly and Baron 1997). However, since tie creation/removal processes alter the structure of the network in which the individual participates, we may wonder whether changes in the topology of the local network surrounding a tie can tell us something about the fact that a new connection is going to appear or has just been removed. In particular, what we want to answer here is: to what extent individuals' choice to create a new tie is related to the their social network before the new connection has been created? Does the topology of their network change after the establishment of a new tie? The same questions can be applied when a connection is removed, instead than created.

The problem we want to study can be formulated as follows. As we have seen, the criterium described in Sect. 3.2.2 allows us to distinguish, for each user, between different types of ties: ties that are active during the whole Ω (before and after), ties that are created within Ω and ties that are removed within Ω. Of course, the two latter conditions can also happen together leading to a fourth type of tie. However, for the purpose of this section, this case will be incorporated in the previous ones, as we will explain soon. To address the questions mentioned above, for each of these categories of ties we study the evolution of the topological overlap (see Sect. 2.2.2) in a time window which spans a period going from before to after the tie has been created or removed. There are at least two reasons why the topological overlap constitutes a good measure to our purpose. First, it gives a measure of the triadic closure (Easley and Kleinberg 2010) which, as mentioned above, is one of the more accepted mechanisms of tie formation. Thus, for example, if two individuals tie together because they have a common friend, we would observe an increase in their topological overlap. In addition, the overlap constitutes a good measure to capture any topological change in the network of a tie also before and after its creation or destruction. In fact, in contrast to other parameters as the tie weight, it can be computed also when the connection does not exist.

To address this, we first consider all those connections that, according to our criterium (see Sect. 3.2.2), are created (removed) within the observation window Ω, i.e. are not observed in the 6-months before (type (b) in Fig. 3.7) or in the 6-months period after (type (a) in Fig. 3.7), respectively. For each tie ij created within Ω, we then focus on the tie's network and analyze the topological overlap O_{ij} across a time window ΔT which spans the period from 100 days before to 50 days after the day in which the tie has formed. In order to ensure the existence of the tie across ΔT, we only consider those ties that are opened in the whole period $\Delta T = 150$ days, with $\Delta T \in \Omega$. We do the same analysis for those ties that are removed within Ω, with the difference that now the overlap is measured from 50 days before to 100 days after the day in which the tie decay is observed. Also in this case, we only consider ties that are opened during the whole period $\Delta T = 150$ days under consideration, with $\Delta T \in \Omega$.

A very known result in social network theory related to the topological overlap is the Granovetter's effect, also known as the "strength of weak ties hypothesis" (Granovetter 1973). Accordingly to this hypothesis, ties between individuals who have many common friends (large overlap), are stronger than the ones between people which few common friends (small overlap), who instead act as bridges between different tight groups (see Chap. 2). This hypothesis has been recently tested for a mobile phone network by observing a positive correlation between O_{ij} and w_{ij}, where w_{ij} measures the strength or weight of the tie and it is given by the total number of phone call communications between i and j (Onnela et al. 2007). In these studies, however, O_{ij} has been calculated on the aggregated network, as well as w_{ij}. Since our analysis allows us to assess with high precision when a tie has been created or removed and, at the same time, to analyze the instantaneous contact network, we take advantage of this and try to investigate also what we call the *dynamical Granovetter effect*. Specifically, in analyzing the evolution of O_{ij} before (after) a connection has been established (removed), we separate all ties in our dataset in different groups according to their weights w_{ij}, and study the evolution of the average $\langle O_{ij} \rangle$ for each group. We calculate the weight w_{ij} as the total number of calls during the whole period of 19 months. This allows us to investigate what comes first in the Granovetter's effect: do people form a large overlap with strong ties or ties that are strong develop a large overlap? Our results are shown in Fig. 3.18 for both ties (a) created and (b) destroyed within Ω. For comparison, we also show the average overlap between pairs of nodes randomly chosen from the whole population, and the average overlap of ties that do not form or decay within the observation window Ω (dashed lines). Several conclusions can be drawn so far. According to the weak ties hypothesis, we observe that topological overlap is strictly related to the intensity or weight of a social relation, meaning that the stronger is the relationship between

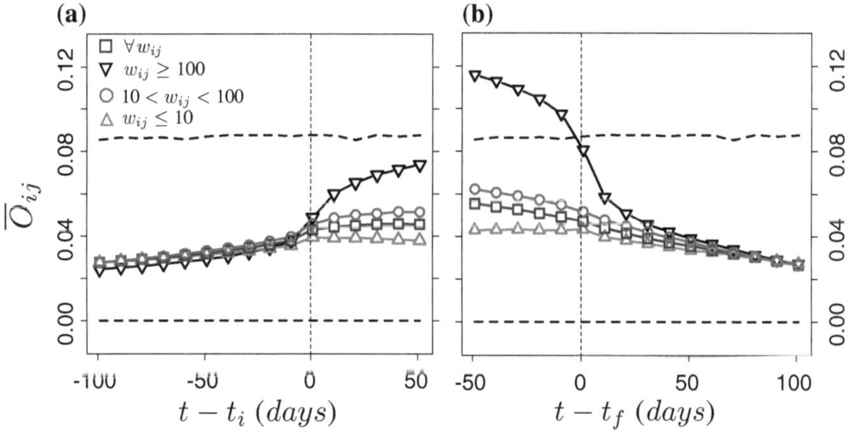

Fig. 3.18 Temporal evolution of the topological overlap O_{ij} averaged over all edges with a given weight w_{ij} within a period of 150 days **a** before the connection has been established and **b** after it has been removed. The time window has been rescaled according to the time at which the tie forms/decays. For comparison, we also plot the average overlap between pairs of nodes randomly chosen from the whole population (*lower dashed line*) and between ties that persist across the whole observation time window Ω (*upper dashed lines*)

two persons, the more friends they have in common, in line with previous results (Granovetter 1973; Onnela et al. 2007).

Moreover, the overlap between two individuals who form (remove) a connection at some time during their lifetime, is significantly higher than the one observed between any random pair of individuals in the whole population even 100 days before (after) the link has been established (removed). This is true also for very weak ties with less than 10 communication events in 19 months. This constitutes a more clear evidence of why the topological overlap is usually a very good feature in the prediction of tie creation (Hidalgo and Rodriguez-Sickert 2008; Liben-Nowell and Kleinberg 2007; Raeder et al. 2011), as we will discuss in the next chapter.
More interestingly, we find that the process that drives two persons to link together is highly dynamical and locally, it entails the change of the underlying topological structure of the network. As shown in Fig. 3.18, in fact, a large value of O_{ij} is observed even before the connection has been established, then it continuously increases over time until the link has been formed and keeps increasing after. The same applies when a connection is removed: the topological overlap between two individuals starts to decrease much before the connection is removed and it keeps decreasing in time after the breakage, although very slowly. This allows us to reply to the question addressed above: what comes first, tie strength or common neighbors? As shown in Fig. 3.18a, what we observe is that, regardless to w_{ij}, the value of O_{ij} is almost the same before the connection has formed and only once the connection is established, it reaches larger values for stronger ties. This result indicates that the connection comes first, and only after the correlation between tie strength and topological overlap starts to form, suggesting indeed a sort of *dynamical Granovetter effect* that, to the best of our knowledge, has not been investigated before.

Our findings also have important practical applications. On the one hand, the fact that the formation of a new tie between two persons can be anticipated (more than) 100 days before by just looking at the number of their common friends is very useful in all those business practices that involves reaching out or developing new relationships as for example, the acquisition of new customers in telecommunication industry. On the other hand, they also have outstanding interest from a sociological and anthropological point of view since they shed more light on the way in which humans establish and remove social connections. We have seen, in fact, that the Granovetter effect is not just a correlation observed in the aggregated contact network, but a dynamical process that happens at a very slow time scale. The correlation between the number of common friends between two individuals and their strength is in fact observable within a time period significantly longer than the lifetime of the social relationship and acts as a *fingerprint* of the social relation itself.

3.6 Discussion

We have analyzed the problem of how people distribute their time and manage their social relations according to the size of their network and the intensity of mobile use. To address this, we first studied the *static* social strategies by analyzing the aggregated social network over a period of 11 months. We thus investigated the role of the temporal patterns of person-to-person communication and the dynamics of tie evolution, which allowed us to characterize how the ongoing temporal changes in human interactions affect the way in which people distribute their time and allocate tasks (*temporal* social strategies).

Our first goal was to understand how people distribute their limited time across their social circle according to their social connectivity and volume of communication. These are the two key variables that characterize the structural topology of an individual's network thus, potentially, they can constitute his first constraint when dealing with the problem of time allocation. Our main results can be summarized in three main points. First, we observe that people with a large personal network spend more time on the phone than people with a small network, which is in line with previous studies. However, in correspondence with a threshold value of the size of the contacts network, the total time people can devote to phone communication reaches a maximum. This indicates that a very large number of contacts does not imply a proportional increase in the amount of time invested in communication. Second, we found that the average time people dedicate to their contacts depends on their network size. In particular, for users with a relatively small number of connections, the time they dedicate to each one of the connections grows proportionally with the network size. However, there is a decrease in the strength of ties for those users with approximately more than 40 connections. This finding is in line with the Dunbar's theory that, up to our knowledge has not been thoroughly investigated for phone interaction networks. Nevertheless, we found that in mobile networks this limit is smaller than the Dunbar number of 150–200 contacts observed for both face-to-face interaction and online social networks, probably because beside the cognitive limit, also temporal and monetary constraints play their role in phone communication. Finally, we studied the user's diversity in time allocation. Interestingly, we found that people do not appear to fundamentally alter the way in which they distribute their time across their network in a way that is related to their intensity of mobile use. All users, no matter whether they have few or many connections, distribute their limited time very unevenly across their network, devoting a large part of their time to a small number of contacts. This is in line with a broad range of findings on both face-to-face and online human communication and show that, whilst mobile phones offer the technical capability to contact everyone in a user's personal network with equal ease, in fact users still focus the majority of their time on a very limited number of contacts.

We also analyzed how the continuous formation and decay of social relationships affect the instantaneous strategies of time allocation. To accomplish this task, we first developed a method to disentangle tie communication activity from tie formation/decay, which is currently one of the main difficulties that hampers a deep understanding of dynamical social strategies. We showed that the way in which people form and destroy communication ties (their social activity) is constant in time and, surprisingly, we found that those processes happen at equal rate, which yields to a steady number of relationships maintained by humans over time (their social capacity). Therefore, although the social network changes rapidly, humans manage their social interactions according to a sort of conservation principle, where the number of outgoing relations equals the number of ingoing ones, a conservation that does not happen only at a particular time scale, but also instantaneously. This result allowed us to characterize individual's social dynamics by means of the unbalance between his *activity* and his *capacity*. In terms of such unbalance, we were able to characterize users with different social strategies of communication and, in particular, to differentiate between users who mostly focus on their social circle (social keepers) and those who, instead, prefer navigate new zones of the network (social wanderers). We found that the adoption of one of those strategies highly correlates with both the topology and the persistence in time of each individuals social neighborhood: while social keepers keep a very stable social circle, social wanderers activate and deactivate a a large number of connections with respect to the one they maintain, resulting in a very volatile social network. In addition, although social strategies tend to not change over the observation time period (thus at a short time scale), they do change across the individual life course. In particular we observe that while young people are characterized by large values of connectivity and high capacity to maintain open relationships, as people get older their social capacity significantly decreases indicating that they maintain a number of stable social contacts much smaller than the one they can establish.

Interestingly, we showed that social strategies are high assortative indicating that social wanderers (as well as social keepers) tend to gather together. This phenomenon has important implications on the global picture of the network, which turns out to be constitutes by static zones of social keepers separated by very volatile zones of social wanderers. We have also seen that the assortative nature of dynamical social strategies reflects in the process of information achievement and spreading: although social wanderers can explore different parts of the network because of their ability to maintain a large social activity, they exhibit a relatively exhibit longer time to receive information spreading in the network compared to social keepers. Therefore, contrary to the belief that exploratory individuals are favored in the process of information access, the fact that they tend to cluster together and have very volatile social circles, actually make them more inefficient.

To shed more light on the dynamic mechanisms of tie creation and removal, we finally analyzed how the network of common friends between two individuals forming a tie, changes before the connection has been established or after it has been removed. To this end, for all those ties that have been created or destroyed within the observation period, we considered different groups of ties accordingly to their weight (measured as the total number of phone calls) and investigated the temporal evolution of their topological overlap before, while and after the connection is active. This study led us to investigate dynamically the *Granovetter effect*, that is the correlation between the strength of a tie and the number of common friends between the two individuals involved. This approach allows to characterize phenomena such as the *strength of weak ties hypothesis* or the *triadic closure mechanism* as dynamical processes and not just as correlations between aggregated quantities. The Granovetter effect, in fact, has been usually addressed from a static perspective, where the above mentioned correlation has been measured on the aggregated contact network. Our analysis, however, allows us to study the *dynamical* process behind such a correlation, since the evolution of the number of common contacts between two individuals is analyzed from before (and after) the strength of their connection emerges. Indeed, we show that the correlation between the strength of a tie and their topological overlap starts to appear only after the social connection has been established. In contrast, it is quite negligible before its formation, although larger than the one observed for any random pair of individual in the whole network.

References

Abramowitz M, Stegun I (1972) Handbook of mathematical functions: with formulas, graphs, and mathematical tables. Dover Publications, New York
Adamic L, Adar E (2003) Friends and neighbors on the web. Soc Netw 25(3):211–230
Adar E, Zhang L, Adamic L (2004) Implicit structure and the dynamics of blogspace. In: Workshop on the weblogging ecosystem
Akoglu L, Dalvi B (2010) Structure, tie persistence and event detection in large phone and SMS networks. In: MLG '10 Proceedings of the eighth workshop on mining and learning with graphs
Almaas E, Kovacs B, Viscek T, Oltvai Z, Barabási A-L (2004) Global organization of metabolic fluxes in the bacterium *Escherichia coli*. Nature 427:839
Anderson RM, May R (1992) Infectious diseases in humans. Oxford University Press, Oxford
Aral S, Van Alsyne M (2007) Network structure and information advantage. In: Proceeding of the academy of management conference, Philadelphia
Bakshy E, Rosenn I (2012) The role of social networks in information diffusion. arXiv:1201.4145v2
Barabási A-L (2010) Bursts: the hidden pattern behind everything we do. Dutton Books, New York
Barabási A-L, Albert R (1999) Emergence of scaling in random networks. Science 286:509–512
Barabási A-L (2005) The origin of bursts and heavy tails in human dynamics. Nat Sci Rep 435:207–211
Barrat A, Barthélemy M, Pastor-Satorras R, Vespignani A (2004) The architecture of complex weighted networks. Proc Natl Acad Sci U S A 101:3747
Barthélemy M, Gondran B, Guichard E (2003) Spatial structure of the internet traffic. Phys A 319:633642
Barthélemy M, Barrat A, Pastor-Satorras R, Vespignani A (2005) Characterization and modelling of weighted networks. Phys A 346:34–43
Becker GS (1965) A theory of the allocation of time. Econ J 75:493–517
Benner M, Tushman M (1999) Process management and organizational adaptation: the productivity dilemma. Unpublished manuscript, Harvard Business School, MA, Boston
Bliss C, Kloumann I, Harris KD, abd Danforth C, Dodds P (2012) Twitter reciprocal reply networks exhibit assortativity with respect to happiness. J Comput Sci 3:388–397
Blumenstock J, Eagle N (2010) Mobile divides: gender, socioeconomic status, and mobile phone use in Rwanda. In: Proceedings of the 4th international conference on information and communication technologies and development
Boccaletti S, Latora V, Moreno Y, Chavez M, Hwang D-U (2006) Complex networks: structure and dynamics. Phys Rep 424:175
Breuer L, Baum D (2005) An introduction to queueing theory. Springer, New York
Burt R (2001b) Structural holes versus network closure as social capital. In: Lin N, Cook K, Burt RS (eds) Social capital: theory and research. sociology and economics: controversy and integration series. Aldine de Gruyter, New York
Burt R (1992) Structural holes: the social structure of competition. Harvard University Press, Cambridge

Burt R (2000) Decay functions. Soc Netw 22:1–28
Carstensen LL (1991) Socioemotional selectivity theory: social activity in life-span context. Annu Rev Gerontol Geriatr 11:195–217
Carstensen LL, Isaacowitz DM, Charles ST (1999) Taking time seriously: a theory of socioemotional selectivity. Am Psychol 54:165–181
Christakis N, Fowler J (2007) The spread of obesity in a large social network over 32 years. N Engl J Med 357:370–379
Christakis NA, Fowler J (2008) The collective dynamics of smoking in a large social network. N Engl J Med 358:2249–2258
Coleman J (1988a) Free riders and zealots: the role of social networks. Sociol Theory 6:52–57
Coleman J (1988b) Social capital in the creation of human capital. Am J Sociol 94:95–120
Coleman J (1990) Foundations of social theory. Belknap Press of Harvard University Press, Cambridge
Cox D (1970) Renewal theory. Methuen and Co, London
Daley D, Kendall D (1964) Epidemics and rumors. Nature 204:1118
Dunbar R (1992) Neocortex size as a constraint on group size in primated. J Hum Evol 22:469
Dunbar R (1998) The social brain hypothesis. Evol Anthropol 6(5):178–190
Eagle N, Macy M, Claxton R (2010) Network diversity and economic development. Science 328:5981
Easley D, Kleinberg J (2010) Networks, crowds, and markets: reasoning about a highly connected world. Cambridge University Press, New York
Ferrara E. (2011). A large-scale community structure analysis in Facebook. arXiv:1106.2503v3.
Frías-Martínez V, Frías-Martínez E, Oliver N (2010) A gender centric analysis of calling behavior in a developing economy using call detail records. In: Proceedings of AAAI artificial intelligence for development
Friedkin N, Johnsen E (1990) Social influence and opinions. J Math Sociol 15:193–206
Gaito S, Zignani M, Rossi G, Sala A, Wang X, Zheng H, Zhao B, (2012) On the bursty evolution of online social networks. In: HotSocial, (2012) first ACM international workshop on hot topics on interdisciplinary social networks research, Beijing, China
Gargiulo M, Benassi M (2000) Trapped in your own net? Network cohesion, structural holes, and the adaptation of social capital. Organ Sci 11:183–196
Getoor L, Diehl CP (2005) Link mining: a survey. ACM SIGKDD Explor Newslett 7:3–12
Ghez GR, Becker GS (1975) The allocation of time and goods over the life cycle. Columbia University Press, New York
Goffman W, Newill V (1964) Generalization of epidemic theory: an application to the transmission of ideas. Nature 204:225
Gonçalves B, Perra N, Vespignani A (2011) Modeling users' activity on twitter networks: validation of Dunbar's number. PLoS ONE 6(8):e22656
Grabowicz P, Ramasco J, Moro E, Pujol J, Eguiluz V (2012) Social features of online networks: the strength of intermediary ties in online social media. PLoS ONE 7(1):e29358
De Graff N, Flap H (1988) With a little help from my friends. Soc Forces 67:453–472
Granovetter M (1973) The strength of weak ties. Am J Sociol 78:1360–1380
Hansen M, Podolny J, Pfeffer J (2001) So many ties, so little time: a task contingency perspective on the value of social capital in organizations. Res Sociol Organ 18:21–57
Hardin G (1968) The tragedy of the commons. Science 162:1243–1248
Hidalgo C, Rodriguez-Sickert C (2008) The dynamics of a mobile phone network. Phys A 387:3017
Hill R, Dunbar R (2003) Social network size in humans. Hum Nat Interdisc Biosoc Perspect 14(1):53–72
Ho V, Rousseau D, Levesque L (2006) Social networks and the psychological contract: structural holes, cohesive ties, and beliefs regarding employer obligations. Hum Relat 59:459–481
Huberman B, Romero D, Wu F (2009) Social networks that matter: twitter under the microscope. First Monday 14:1

Ibarra H (2002) Homophily differential returns: sex differences in network structure and access in an advertising firm. Adm Sci Q 37:422

Kahn R, Antonucci T (1980) Convoys over the life course: attachment, roles and social support. In: Baltes P, Brim O (eds) Life-span development and behavior. Academic Press, New York

Karsai M, Kivelä M, Pan R, Kaski K, Kertész J, Barabási A-L (2011) Small but slow world: how network topology and burstiness slow down spreading. Phys Rev E 83:025102(R)

Kikas R, Dumas M, Karsai M (2012) Bursty egocentric network evolution in Skype. arXiv:1206.0883v1

Kossinets G, Watts DJ (2006) Empirical analysis of an evolving social network. Science 311:5757

Krings G, Karsai M, Bernharsson S, Blondel V, Saramäki J (2012) Effects of time window size and placement on the structure of aggregated networks. EPJ Data Sci 1:1–4

Lee S-H, Kim P-J, Ahn Y-Y, Jeong H (2010) Googling social interactions: web search engine based social network construction. PLoS ONE 5(7):e11233

Leskovec J, Lang KJ, Dasgupta A, Mahoney M (2009) Community structure in large networks: natural cluster sizes and the absence of large well-defined clusters. Internet Math 61(1):29–123

Leskovec J, Backstrom L, Kleinberg J (2009) Meme-tracking and the dynamics of the news cycle. In: KDD '09 Proceedings of the 15th ACM SIGKDD international conference on knowledge discovery and data mining, pp 497–506

Leskovec J, Backstrom L, Kumar R, Tomkins A (2008) Microscopic evolution of social networks. In: KDD '08 Proceedings of the 14th ACM SIGKDD international conference on knowledge discovery and data mining, pp 462–470

Liben-Nowell D, Kleinberg J (2007) The link-prediction problem for social networks. In: CIKM '03: Proceedings of the twelfth international conference on information and knowledge management, New York, NY, USA, pp 556–559

Linder SB (1960) The Harried Leisure class. Columbia University Press, New York

March J (1991) Exploration and exploitation in organizational learning. Organ Sci 2:71–87

Marlow C (2009) Maintained relationships on Facebook. http://overstated.net

McPherson J, Smith-Lovin L, Cook J (2001) Birds of a feather: homophily in social networks. Annu Rev Sociol 27:415–444

Miritello G, Lara R, Cebrián M, Moro E (2012a) Limited communication capacity unveils strategies for human interaction. arXiv:1304.1979

Miritello G, Lara R, Moro E (2013) Time allocation in social networks: correlation between social structure and human communication dynamics. In: Holme P, Saramäki J (eds) Temporal networks. Springer, Berlin

Miritello G, Moro E, Lara R (2011) Dynamical strength of social ties in information spreading. Phys Rev E 83:045102(R)

Miritello G, Moro E, Lara R, Martinez R, Belchamber J, Roberts S, Dunbar R (2012) Communication strategy in mobile phone networks. Submitted to Social Networks, Time as a limited resource

Mollica K, Gray B, Trevino L (2003) Racial homophily and its persistence in newcomers' social networks. Organ Sci 14:123–136

De Montis A, Barthélemy M, Chessa A, Vespignani A (2007) The structure of interurban traffic: a weighted network analysis. Environ Plan B Plan Des 34(5):905–924

Morgan D, March SI (1992) The impact of life events on network of personal relationships: a comparison of widowhood and caring for a souse with Alzheimer's disease. J Soc Pers Relat 9:563–584

Murray J (1993) Mathematical biology. Springer, Berlin

Newman MEJ (2002a) Assortative mixing in networks. Phys Rev E 89:208701

Newman MEJ (2003b) The structure and function of complex networks. SIAM Rev 45:167–256

Onnela J-P, Saramäki J, Hyvönen J, Szabó Z, Lazer D, Kaski K, Kertész J, Barabási A-L (2007) Structure and tie strengths in mobile communication networks. Proc Natl Acad Sci U S A 104:7332

Palchykov V, Kaski K, Kertész J, Barabási A-L, Dunbar R (2012) Sex differences in intimate relationships. arXiv:1201.5722v2

Palla G, Barabási A-L, Tamás V (2007) Community dynamics in social networks. Noise Stochast Complex Syst Finance 6601:660106

Podonly J, Baron J (1997) Resources and relationships: social networks and mobility in the workplace. Am Sociol Rev 62:673–693

Putnam R (1993) The prosperous community: social capital and public life. Am Prospect 13:35–42

Quattrocchi W, Conte R, Lodi E (2010) Simulating opinion dynamics in heterogeneous communication systems. In: Proceedings of 7th European conference on complex systems (ECCS), pp 70–85

Raeder T, Lizardo O, Chawla N, Hachen D (2011) Predictors of short-term decay of cell phone contacts in a large scale communication network. Soc Netw 33:245–257

Rapoport A (1953) Spread of information through a population with socio-structural bias: I. assumption of transitivity. Bull Math Biophys 15:523–533

Roberts S (2010) Constraints on social networks. Proc Br Acad 158:115–134

Rogers E (1995) Diffusionon of innovations. Free Press, New York

Roth M, Ben-David A, Deutscher D, Flysher G, Horn I, Leichtberg A, Leiser N, Matias Y, Merom R (2010) Suggesting friends using the implicit social graph. In: KDD '10 Proceedings of the 16th ACM SIGKDD international conference on knowledge discovery and data mining, pp 233–242

Saramäki J, Leicht E, Lopez E, Roberts S, Reed-Tsochas F, Dunbar R (2012) The persistence of social signature in human, communication. arXiv:1204.5602

Seeman T, Miller-Martinez D, Stein-Merkin S, Lachman M, Tun P, Karlamangla A (2012) Histories of social engagement and adult cognition in middle and late life: the midlife in the u.s. study. J Gerontol Psychol Sci 66B:141–152

Seibert S, Kraimer M, Linden R (2001) A social capital theory of career success. Acad Manag J 44:219–237

Southerton D (1986) Squeezing time: allocating practices, coordinating networks and scheduling society. Time Soc 12:5–25

Suitor J, Wellman B, Morgan D (1997) It's about time: how, why, and when networks change. Soc Netw 19:1–7

Sutcliffe A, Dunbar R, Binder J, Arrow H (2012) Relationships and the social brain: integrating psychological and evolutionary perspectives. Br J Psychol 103:149–168

Tantipathananand C, Berger-Wolf T, Kempe D (2007) A framework for community identification in dynamic social networks. In: Proceeding KDD '07 Proceedings of the 13th ACM SIGKDD international conference on knowledge discovery and data mining, pp 717–726

Tilburg T (1998) Losing and gaining in old age: changes in personal network size and social. Support in a four-year longitudinal study. J Gerontol Soc Sci 53B:313–323

Ugander J, Backstrom L, Marlow C, Kleinberg J (2012) Structural diversity in social contagion. Proc Natl Acad Sci U S A 109(16):5962–5966

Vázquez A, Oliveira J, Dezsö Z, Goh K-I, Kondor I, Barabási A-L (2006) Modeling bursts and heavy tails in human dynamics. Phys Rev E 73:036127

Vázquez A, Rácz B, Lukács A, Barabási A-L (2007) Impact of non-poissonian activity patterns on spreading processes. Phys Rev Lett 98:158702

Wellman B, Wortley S (1990) Different strokes from different folks: community ties and social support. Am J Sociol 96:558–588

Wellman B (2007) The network is personal: introduction to a special issue of social networks. Soc Netw 29(3):349–356

Winship C (1978) The allocation of time among individuals. In: Schuessler K (ed) Sociological methodology. Jossey-Bass, San Francisco

Wrzus C, Hänel M, Wagner J, Neyer F (2012) Social network changes and life events across the life span: a meta-analysis. Psychol Bull 10:a0028601

Zhao Q, Oliver N (2010) Communication motifs: a novel approach to characterize mobile communications. In: NetMob2010

Chapter 4
Predicting Tie Creation and Decay

Other things equal, relationships can be expected to weaken over time such that some observed today are gone tomorrow.
—Ronald S. Burt

As established in the previous chapter, social networks are dynamic objects, they grow and change over time through the addition of new ties or the removal of old ones, leading to an ongoing appearance and disappearance of interactions in the underlying social structure. Identifying the mechanisms by which a link form or decay is a fundamental and challenging question not only in social network analysis, but in many data mining tasks. Example are abundant. Classical information retrieval can be viewed as predicting missing links between words and documents (Craven et al. 2000; Salton and McGill 1983), items recommendation to a user can be considered as a link prediction problem in the user-item bipartite network (Zhou et al. 2007). Detection of future links also helps in predicting participation of actors in events (O'Madadhain et al. 2005) or in the friend-suggestion mechanisms used in many online social networks (Roth et al. 2010). In general, there are two ways in which the prediction of link occurrence can be utilized. The first applies to all those situations where not all links are observed, in which cases one may be interested in inferring missing (or anomalous) links from the observed network. The other, instead, refers to situations in which links evolve over time, in which case the goal is to predict whether a link will exist in the future, based on the observation in a given time window. When dealing with social networks, the interest in studying the problem of link prediction goes beyond the specific interest of predicting what ties are more likely to persist or decay in the future. As discussed in the previous chapter, in many cases there is no explicit declaration for the existence of a tie, which is instead inferred from tie activity. However, tie activity does not necessarily imply the existence of a very social connection (Butts 2009). For this reason in social networks such as e-mail or cell phone networks, the concept of what exactly constitutes an edge is indeed somewhat

"Attachment, Decay, and Social Network" (Burt 2001a).

G. Miritello, *Temporal Patterns of Communication in Social Networks*, Springer Theses, DOI: 10.1007/978-3-319-00110-4_4,

unclear (Hidalgo and Rodriguez-Sickert 2008; Kossinets and Watts 2006) since a communication event between two individuals may refer to a spam message or to a wrong-number call. For this reason, understanding under what condition a tie is more or less likely to decay may shed light on the circumstances under which an observed tie can be actually considered a very social relationship. In addition, the fact that a given individual shows a high level of volatility and decay in his current ties may also indicate that he is moving from one community to another (Suitor et al. 1997). In this sense, tie decay and creation could signal community structure changing or formation (Palla et al. 2007; Tantipathananand et al. 2007), with important implications at the level of the whole network. As a consequence of these observations, the problem of link prediction is strictly related to the task of how to model dynamical social networks, which, as mentioned in the previous chapters, remains a understudied topic (Liben-Nowell and Kleinberg 2007). In fact, although the literature about the problem of link prediction is vast and growing (Getoor and Diehl 2005), this issue is yet not well understood; for example, to what extent structural features of social relationships allow to predict and model the evolution of a social network?

Of course, there are multiple forces that govern the formation and the removal of a social tie between individuals, both endogenous and exogenous. However, as we will discuss in this cphapter, it has been observed that endogenous factors, i.e. those properties that can be extrapolated from the network itself, are very good predictors for tie creation and decay. They reflect, in fact, the inherent mechanisms that lead people to tie together and/or maintain a social relationships, such as triadic closure (Burt 1992; Easley and Kleinberg 2010; Granovetter 1973) or homophily (McPherson et al. 2001). For this reason, the majority of approaches to this issue focus to identify tie properties such as common friends or communities properties (Hidalgo and Rodriguez-Sickert 2008; Kossinets and Watts 2006; Raeder et al. 2011), similarity among individuals (Adamic and Adar 2003; Aral et al. 2009; Crandall et al. 2008; Kossinets and Watts 2009) or volume of interaction, which have been observed to capture the emotional intensity of a tie (Hill and Dunbar 2003; Saramäki et al. 2012; Wellman and Wortley 1990). Another conclusion that emerges from previous studies is that ties exhibit a memory, in the sense that old ties are more likely to persist in time than newly-formed ones (Burt 2000; Raeder et al. 2011).

However, what it is usually neglected in all these studies is the information about tie occurrences and the temporal rhythms of tie interaction. Only very recently, in fact, time-dependent features related to the temporal evolution of dyadic communication have been taken into consideration in this area (Gilbert and Karahalios 2009; Raeder et al. 2011). According to these findings, the prediction in tie short-term decay can be improved by taking into account edge longevity (time from the first communication) and edge freshness (time since the last communication). However, these studies have been conducted over relatively short time windows (order of months), in which cases, as we have seen in the previous chapter, the appearance and decay of a social relationship can be confused with the (temporal) tie inactivity (Miritello et al. 2012). A small variation of the period under investigation may therefore dramatically affect the results.

In line with what observed in the previous chapter for an individual's network, here we show that, actually, the bursty human interaction leads to a misleading picture of the global network, by making the observed tie decay process much faster than the real one. To separate tie activity from tie creation and removal in a given time window, in the previous chapter we have proposed a method based on the very observation of ties before and after the observation period. In order for the method to be efficient, however, long time periods are required, which is usually a hard task to accomplish. Here we introduce a simple method that allows us to infer with good precision whether a social connection, observed in a given time period, can be considered a newly-formed (decayed) tie or whether it exists from before (persist after) the observation window. Although its main purpose lies in the analysis of the link prediction problem in social network, our study can be therefore used also to address the analysis done in Sect. 3.3 of social strategies of communication in all those cases where a short observation period is available.

As mentioned above, we are interested in analyzing the role that human activity patterns plays in the link prediction problem. For this reason we first characterize the temporal properties of communication ties of a large mobile phone network and propose some simple time-dependent features that have not been used in previous research to characterize the strength of a tie. The analysis of these features allows to distinguish between ties characterized by the same topological properties, but with very different activity patterns. Based on such temporal properties, we then introduce and analyze a simple method to predict tie persistence and decay and show that it gives better results than more complex standard models mainly based on topological aspects of the network. Before presenting our analysis, however, we briefly review the standard methods used to address the link prediction problem in social networks and discuss the main results present in the literature.

Our study constitutes one of the first steps towards the understanding of how temporal properties of human communication can be used (together with structural, geographical or homophily-driven factors) in the prediction of which connections are more likely to be created or destroyed in the near future. More importantly, it also helps in the more general and challenging problems of social network analysis: the identification and characterization of the social relationship behind an observed activity pattern.

4.1 Conventional Approaches to the Link Prediction Problem

Most of the problems related to the prediction of the existence of links among nodes can be described in terms of the so-called *link prediction problem*, i.e. the estimation of the probability that a link will emerge or disappear during a future time window. According to the general formulation given by Liben-Nowell and Kleinberg (2007), the link prediction problem can be formulated as follows: given a graph $\mathcal{G} = \{\mathcal{P}, \mathcal{E}\}$ (see Sect. 2.1.1) at time t, predict which (or with what probability) new ties will form between the vertices $\mathcal{P}$ in the time interval $t + \Delta t$. A similar problem can be formulated to predict tie decay in a future time window, instead than tie formation. The conventional way of addressing this problem is based on the assumption that many real networks are endogenous, thus their properties at a given time contain information on the status of the networks in the future. Although a number of factors, both endogenous and exogenous usually contribute towards tie creation and decay in social networks, there is a vast literature suggesting that information about future interactions can be actually extracted from the network structure (Aiello et al. 2010; Akoglu and Dalvi 2010; Eagle et al. 2009; Hidalgo and Rodriguez-Sickert 2008; Liben-Nowell and Kleinberg 2007; Kossinets and Watts 2006; Raeder et al. 2011).

To address this, one usually splits the time period under investigation into two subsequent windows τ_1 and τ_2, thus obtaining two snapshots of the same network. Having characterizing nodes and ties of the network observed in τ_1 by a set of features, one then tries to build a generalizable model to predict, based on those features, the tie persistence or the decay in τ_2. This corresponds to assign to each of the ties observed in τ_1 what is called a *predictive class*, which in many cases is a categorical variable taking the value 0 if the tie is predicted to decay or not observed in τ_2 (decayed class) and 1 otherwise (persistent class). The predicted class is then tested to be true or false depending on whether the observation of the network in τ_2 validates it or not. For this reason one usually refers to τ_1 and τ_2 as, respectively, *prediction* and *observation* time interval. To ensure the effectiveness of model, the available set of ties is randomly divided into two different subsets within each time period τ_1 and τ_2. One set, called *training data*, is used to build the model, while the other, called *testing data*, to validate it.

To build the prediction model, plenty of machine learning algorithms have been proposed and applied to various real networks (Huang and Lin 2009; Getoor and Diehl 2005) which include, among others, Markov networks (Richardson and Domingos 2006; Sarukkai 2000), probabilistic relational models (Song et al. 2009; Wang et al. 2007), stochastic relational models (Yu et al. 2007) or simple regression models as decision-tree classifier and logistic regression (Burt 2000; Raeder et al. 2011).
A different group of algorithms is instead based on the definition of node similarity or homophily (McPherson et al. 2001), discussed in the previous chapters. According to these models, two nodes are considered more likely to create a connection if they are similar and thus, if they have many common features. Among all the features that in principle can be used to assess nodes' similarity, such as common behavior, common interests or common neighbors, one usually measures quantities related to the latter one and, in particular, local similarity indices such as the Jaccard similarity or Adamic-Adar index (Adamic and Adar 2003; Liben-Nowell and Kleinberg 2007; Lü and Zhou 2009) since they can be easily inferred from the structure of the network. Sometimes, the similarity between the members of a networks is computed by means of the concept of *proximity*: the closest two nodes within a network, the larger the probability they will connect in the future. This class of methods follows the notion that many networks, and especially social networks, are small word (see Sect. 2.1.1) thus agents are connected through short chains. To capture the proximity between pairs of nodes several measures are typically applied, as PageRank (Page et al. 1999), Katz measure (Katz 1953) or SimRank score (Glen and Widom 2002).

Another approach, which is the one we follow in our analysis, consists in a *threshold based* method, where the decayed or persistent class is assigned depending on whether one of the measured features used to characterize tie strength or similarity satisfies or not a certain condition (Hidalgo and Rodriguez-Sickert 2008). Specifically, when the features considered are binary variable, such as tie reciprocity, one usually assigns the persistent or the decayed class depending on whether it is satisfied or not. When, instead, the feature is either a categorical or continuous variable (tie weight or topological overlap), the persistent or decayed class is assigned depending if the value of such a variable is, respectively, greater or smaller than a threshold θ, where the value of θ can be tuned to improve the performance of the model. Despite its simplicity, this approach leads to a successful prediction of tie persistence, as we will see in Sect. 4.4.2.

4.2 Characterizing a Social Tie

As emerges from the previous section, the problem of predicting tie persistence or decay is typically reduced to identify those features that can better characterize a tie. The extant literature on link prediction in social networks has focused mostly in the use of *topological* features, measured within the prediction window τ_1 to infer tie persistence or decay in the observation window τ_2, while *temporal* features of tie communication have been usually neglected. This is a reflection of the traditional way in which social networks are typically modeled. In fact, as discussed in Sect. 2.3, the standard way to describe social networks is based on the study of a static and aggregated picture of the interactions network in the observation window; within this picture, ties are characterized by aggregated quantities which do not take into account the temporal aspects of human communication. However, as we have already seen, although topological features capture many structural aspects and mechanisms of real social networks, temporal patterns of communication are essential to have a more realistic characterization of the contact network. Only very recently, *time-dependent* parameters have been also considered to predict the decay of ties. In this context, the most representative (and perhaps the only one) study is the one by Raeder et al. (2011) where, besides topological features such as tie strength or topological overlap, the authors also consider the time from the first communication (*newness*) and the time since the last communication (*freshness*) of a tie as predictors for the short-term decay of relationships in a mobile phone communication network. However, we have already shown in Sect. 3.2 that when observing a network in a given time window, especially if it is short, the bursty activity of tie interaction can lead to a misleading picture of the contact network. As a consequence, quantities such as the time from (since) the first (last) communication may only reflect the intensity of the tie communication, but they do not capture the temporal properties of how interaction events are allocated in time. Nevertheless, since as well as topological properties, also temporal properties of tie communication contain useful information about the underlying social relationships, we expect that they will significantly improve the current formulation of the link prediction problem in social networks.

Actually, the interest in characterizing social ties by taking into account also the time patterns of human interaction goes beyond the specific problem of predicting tie appearance or decay, since it would lead to a dynamical characterization of the social network, in contrast to the aggregated static one. To address this, in this section we propose some time-dependent variables that, to the best of our knowledge, have not been directly used in previous research to characterize social ties. Firstly, however, we briefly present some of the topological features of social networks typically used in the literature to address the link prediction problem.

4.2.1 Topological Features for Tie Persistence and Decay

As described in Chap. 2, some of the main characteristics of social networks are assortativity, homophily nature and community structure (Newman and Park 2003). This is the main motivation behind the use of topological variables in predicting the short term future of a tie and, at the same time, also the reason for their success, since they are able to capture the main properties of what the mechanism of tie creation and/or stability seems to be. Typically, the topological features used in the link prediction problem to characterize a tie between two individuals can be divided into three main categories. The first category is composed by *vertex attributes*, which generally includes nodes' (in and out) degrees, overall aggregated activity (or strength), usually measured as the total number of communication events and centrality measures as the clustering coefficient. In general, vertex features are positively correlated with the creation of a tie and contribute significantly to the predictive power of the model, a result that reflects assortative mixing and correlations in the degree and the activity discussed in the previous chapters (Aiello et al. 2010; Liben-Nowell and Kleinberg 2007; Raeder et al. 2011). Predictive models significantly improve when taking into account also *dyad-level attributes* (or tie-level attributes) (Lü and Zhou 2010; Raeder et al. 2011; Zhou et al. 2009). Specifically, it has been found that variables such as tie reciprocity and tie weight, reflecting the balance of the relationship and its importance, are two of the most important indicators of tie persistence/decay: the stronger the tie (in terms of volume of communication), the less likely it is to decay over time (Hidalgo and Rodriguez-Sickert 2008; Raeder et al. 2011). Another variable that has been identified to be highly predictive, also due to its positive correlation with tie features such as the tie strength (Granovetter 1973), is the topological overlap, which as mentioned before reflects the level of their similarity (Hidalgo and Rodriguez-Sickert 2008; Zhou et al. 2009). This measure belongs to a third class of topological measures, the *neighborhood-level* or *local features*, since they also involve nodes besides the ones constituting the tie. Generally, these features are not considered alone in predictive methods. Instead, a combination of some (of all) the above mentioned features, among others, is assigned to a tie and used to assess its decayed or persistent class.

4.2.2 Measuring Temporal Features of Social Ties

At this stage it should be clear that temporal aspects of human interaction are crucial to characterize the type of social relationship which is behind an observed activity pattern. As mentioned above, as well as static quantities such as the volume of communication between two individuals, also the way in which such communication is distributed within a given temporal period must reflect the type and the importance of the relationship. While two individuals who interact every day may be tied together by a family relationship or a close friendship, a more irregular communication may refer to an occasional (although close) social tie. The fact that aggregated quantities, such as the total number of interaction events, are not able to differentiate between these types of relationship becomes clear by looking at Fig. 4.1a, where a schematic representation of the communication pattern of different ties is depicted. Each tie is characterized by the same number of calls w_{ij}, usually used to define tie weight; their temporal activity pattern, however, significantly differ from each another, meaning that different patterns of communication can correspond to the same weight. Moreover, note that even quantities such as the time from (since) the first (last) communication, alone, are not enable to capture the observed differences between the different patterns, as for example the burstiness of the communication.

To capture these differences, we introduce and analyze some temporal features that, to the best of our knowledge, have not been previously used to characterize the strength of a social tie (Miritello et al. 2013). First, we define the *temporal stability* Δ_{ij} of a tie as $\Delta_{ij} = t_{ij}^{max} - t_{ij}^{min}$, where t_{ij}^{min} and t_{ij}^{max} are, respectively, the time instants at which the first and the last communication event between i and j are observed within the observation period T (shadowed areas Fig. 4.1a). A large stability $\Delta_{ij} \simeq T$ (cases A, B or C in Fig. 4.1) indicates that the communication between i and j may extend over the observation period, while a small stability

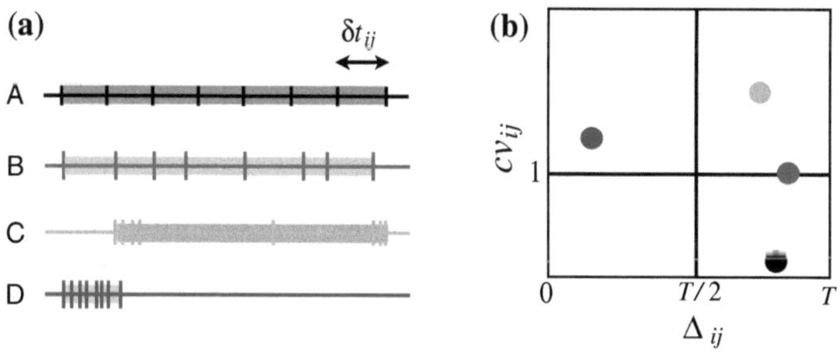

Fig. 4.1 **a** Schematic representation of the temporal activity pattern of different communication ties. Each line refers to a different tie while each vertical segment indicates a communication event within the tie which inter-event time is δ_{ij}. All ties are characterized by the same number of events (same number of calls $w_{ij} = 8$) but very different temporal pattern of communication. **b** Bivariate representation of the coefficient of variation cv_{ij} and the temporal stability Δ_{ij} of ties for the corresponding cases in **a**. Adapted from "Time allocation in social networks: correlation between social structure and human communication dynamics" Miritello et al. (2013)

$\Delta_{ij} \simeq 0$ (case D in Fig. 4.1) is the signal of a short tie communication. It should be noticed that $\Delta_{ij} \simeq T$ defines tie stability (or lifetime) of a tie within a given observation window T. However, as seen in the previous chapter, due to the high burstiness of tie interaction, the fact that a short communication is observed within T does not necessarily imply a shorter relationship. The latter, in fact, is not captured by Δ_{ij}, which does not allow to differentiate between an homogeneous communication tie (case B in Fig. 4.1) and a more heterogeneous one (case C in Fig. 4.1).
Although there are other ways to characterize and define the burstiness of a tie communication (Goh and Barabási 2008; Karsai et al. 2011), to capture this information we use the common *coefficient of variation* cv_{ij} of the inter-event time distribution $P(\delta t_{ij})$ of a tie. It is measured as $cv_{ij} = \sigma_{\delta t_{ij}} / \overline{\delta t}_{ij}$, where $\overline{\delta t}_{ij}$ and $\sigma_{\delta t_{ij}}$ are respectively the mean and the standard deviation of the δt_{ij} series. By definition cv_{ij} measures the dispersion of a distribution, that is the level of heterogeneity of tie communication: the more heterogeneous the interaction, the larger cv_{ij}.
Thus, for a perfect deterministic interaction in which events happen at regular times, as in the case A of Fig. 4.1, we get $cv = 0$, while for a homogeneous Poisson-like process as the case B we have that $cv \simeq 1$ since $\sigma_{\delta t_{ij}} \simeq \overline{\delta t}_{ij}$. In contrast, a bursty communication in which the inter-event time distribution is heavy-tailed (see Sect. 2.4), we have $\sigma_{\delta t_{ij}} \gg \overline{\delta t}_{ij}$, thus $cv \gg 1$ (case C of Fig. 4.1). As showed in Fig. 4.1b, contrary to the number of communication events w_{ij}, a combination of the temporal stability and coefficient of variation allow us to differentiate between different patterns of tie communication.
Note that, as discussed in Sect. 2.3, characterizing social ties only by their weight w_{ij} correspond to a "poissonization" of the real process. In this latter case, since events are evenly distributed across the observation window T, we get that $\Delta_{ij} \simeq T$ and $cv_{ij} \simeq 1$. This is shown in Fig. 4.2, where we show the density plot of both these quantities measured for ties in the mobile phone network described in Appendix A during a period of $T = 11$ months and compare them with the one obtained in the Poisson-like case. To mimic the latter process we have shuffled the real time-stamps of communication events across the whole database, thus each call has the same probability to appear anytime within the observation window (see Appendix A).
As expected, in the Poisson-like case we observe that most of the ties have $\Delta_{ij} \simeq T$ and $cv_{ij} \simeq 1$,which significantly differ from the results obtained in the real case. In the latter case, in fact, most ties show bursty behavior ($cv_{ij} \gg 1$) and a unevenly distributed stability across the observation window. In line with previous studies on mobile phone networks (Hidalgo and Rodriguez-Sickert 2008), we observe that a large proportion of ties have either large or very small stability showing that ties have mostly a very long or a very short lifetime, possibly signaling the very different nature of the communication involved: while long ties might reflect close personal relationships, short ones could be associated to more volatile ties, such as work relationships (Miritello et al. 2013).

This result clearly indicates how important is considering temporal patterns of communication in the description and characterization of a social network. By looking at Fig. 4.2, in fact, it becomes clear how different the picture of social ties

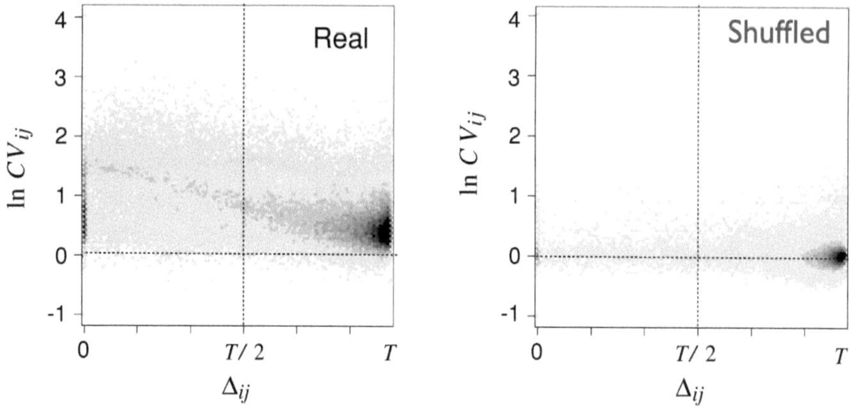

Fig. 4.2 Density plot of the logarithm of the coefficient of variation and stability of the ties in our database for the real sequence of calls (*left*) and the corresponding ones for the shuffled time stamps (*right*). To have enough statistical significance for cv_{ij}, only ties with $w_{ij} \geq 5$ are used in the plot. Adapted from "Time allocation in social networks: correlation between social structure and human communication dynamics" Miritello et al. (2013)

would be from the real one when the temporal dimension is projected out. Finally, in addition to the stability and the coefficient of variation, we also measure the average inter-event time $\overline{\delta t}_{ij}$, given by $\overline{\delta t}_{ij} = \Delta_{ij}/w_{ij}$, where Δ_{ij} is the tie stability and w_{ij} the number of calls. This feature capture the frequency of tie communication and, more importantly, it encompasses all the information about the distribution $P(\delta t_{ij})$ (Candia et al. 2008; Goh and Barabási 2008; Karsai et al. 2011), as we will see in Sect. 4.4.2. Of course, Δ_{ij}, cv_{ij} and $\overline{\delta t}_{ij}$ are correlated to each other as well as to other topological variables: the stability, for example, increases with the number of calls and, for a given stability, the average time decreases as the number of calls increases.

Before analyzing the correlations between tie features, however, we would like to show that the problem of time allocation and communication strategies analyzed in the previous chapter, also emerges with respect of the temporal features of ties. Specifically, we have seen that cognitive and time constraints influence the aggregated amount of attention per tie for individuals with a large social connectivity k_i (Gonçalves et al. 2011; Miritello et al. 2012; Saramäki et al. 2012) (see Sect. 3.1). Since communication events are not evenly distributed across the observation period, we might expect that the fact that some individuals are constrained to allocate time in their social network, also reflects in the way in which they distribute the communication within their connections. For example, for a given w_{ij}, the attention allocated in a short time is much more localized in time than in a long tie. Or individuals might choose to develop more bursty activity with some connections to allocate more conversations within the day. To address this, we analyze how Δ_{ij} and cv_{ij} depend on the social connectivity k_i. Our results are shown in Fig. 4.3 for a small sample of individuals in the phone call database we are analyzing. As one can see, we find

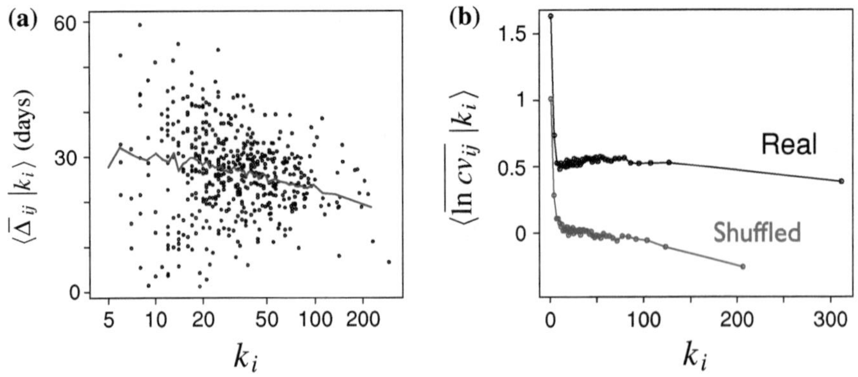

Fig. 4.3 **a** Scatter plot and conditional mean (*red curve*) of the average stability of ties Δ_{ij} for a given individual for a small sample of individuals of the mobile phone database under consideration as a function of the social connectivity k_i. We only shows the results for the real-time sequence of calls, since for the Poissonian reference model $\overline{\Delta}_{ij} \simeq T$. **b** Conditional mean of the average logarithm of the coefficient of variation cv_{ij} for the real-time sequence of calls (*black curve*) and for the Poissonian shuffled one (*red curve*) as a function of the social connectivity k_i Adapted from "Time allocation in social networks: correlation between social structure and human communication dynamics" Miritello et al. (2013)

that while, on average, there is no significant dependence of cv_{ij} on k_i, the larger the social connectivity, the smaller the average stability of ties. This means that, regardless to their number of connections, people always allocate time burstly in time; however, more connected individuals show on average shorter communication ties.

This is a clear indication that, aside from the non trivial way in which people allocate time among their connections, there is also a complex way in which their attention unfolds in time. In Chap. 5 we will see that the observation that more connected people have weaker and shorter communication within their social circle, also affects their transmission power when dealing with information transmission processes (see Sect. 5.3). Taken together, these findings constitute a further evidence of the importance of considering temporal aspects of human activity in the description and modeling of social networks (Miritello et al. 2013).

4.3 Correlations Between Tie Features

When dealing with tie persistence/decay prediction problems, the more features can be extracted from the network, the better. However, due to the strong correlations between variables, most of the information may be redundant and captured by only few of them (Newman and Park 2004; Vázquez et al. 2004). Although correlations reveal important features of individuals properties and behavior, they can however lead to misleading conclusions in classification problems. For this reason, to unveil the real contribution of each metrics to the tie persistence and decay prediction, the analysis of correlations between the predictive variables is usually strongly recommended before applying any type of predictive model. It may help, in fact, in the selection of the subset of relevant variables, which is usually a hard task in data analysis especially when dealing with many variables. This selection usually increases the robustness of models and simplifies the understanding of the outcomes (Lemaire et al. 2009). Figure 4.4 shows the correlations between tie features for the mobile phone network that we are considering. Some of these features, such as the bias, the number of calls, the topological overlap and the Adamic-Adar index, have been introduced in Sect. 2.2.2 and, as mentioned above, they play an important role in the link prediction problem in social networks. As expected, there is a positive (and significant) correlation between variables that identify the strength of a relationship and the number of common friends, in line with previous studies (Granovetter 1973; Kovanen et al. 2011; Onnela et al. 2007).

Another group of features, which instead are time-dependent, includes the time since the last communication (freshness) (Raeder et al. 2011) and the three variables introduced in the previous section: the temporal stability, the coefficient of variation and the average inter-event time. As expected, the latter group of features also correlates with the topological ones. For example, there is a positive correlation between the number of calls and the stability: the longer the lifetime of a tie, the larger the number of communication events.

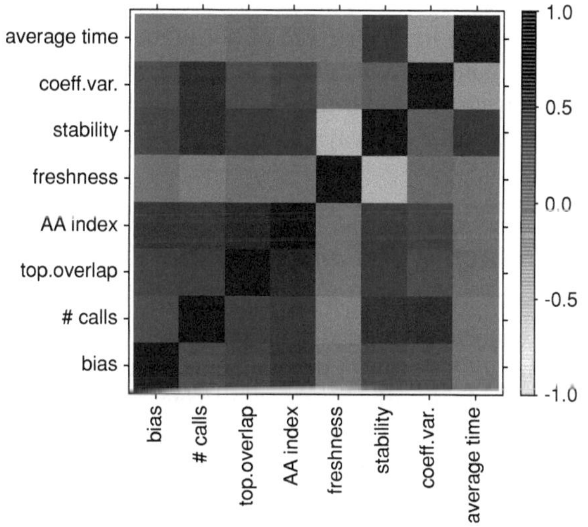

Fig. 4.4 (Pearson) Correlation matrix between some of the main representative (topological and temporal) features of ties observed in our mobile data set

4.4 Temporal Patterns and Tie Persistence/Decay

As discussed in the previous section, temporal aspects of human interaction have been usually neglected in the characterization of a social tie and, as a consequence, also in the tie persistence/decay problem. Nevertheless, they gather important information on the nature of a social relationships, which is fundamental to distinguish between ties with the same topological properties, but very different temporal patters (see Fig. 4.1). In this section we are interested in analyzing whether, as well as topological features, temporal properties of tie communication patterns may help in the prediction of its persistence or decay in the future. Specifically, the question we address is the following: does the way in which two individuals communicate in a given period of time contain information of how likely they are to interact in a future time window? The approach we follow is the one described in Sect. 4.1: we consider two contiguous subintervals of the total time period under investigation and measure the temporal properties of the ties in the prediction time period τ_1. We then use the observation time interval τ_2 to validate the goodness of out model. There is however an important issue, related to the analysis conducted in the previous chapter, that we want to stress before presenting our method. We have seen that the heterogeneity observed in the human communication patterns can have important consequences in the definition of whether a social relationship has actually decayed or not, since the non-observation of activity between two individuals in a given time period can be either associated to a period of inactivity (burst) as well as to the very decay of the tie. This has crucial implications in the problem of link prediction, where the predictive power of each tie feature, as well as the goodness of the model, is actually inferred depending on whether the tie is observed or not in the prediction time window. For this reason, before going into the details of our model, we briefly address the latter issue and investigate the implications of the bursty human communication in the definition of tie persistence or decay.

4.4.1 Active Ties and Open Ties

The analysis performed in the previous chapter has shown that, as a consequence of the bursty behavior of tie communication, the measured rate of tie decay within an individual network significantly changes from the real one when the existence of a tie is assessed only on the basis of its activity (see Fig. 3.8). This led us to distinguish between observed ties and open (or active) ties: an observed tie is a tie that shows at least one interaction within a given time period T, while an open tie is a tie that, whether or not it interacts within T, it is observed both before and after.

The difference between the two definitions becomes more clear by looking at the schematic illustration in Fig. 4.5, where we show four consecutive snapshots of a schematic network of five nodes. If we focus on the panel $T = 2$, the tie $1 \leftrightarrow 4$ between nodes 1 and 4 is an open relationship since, although it does not show any

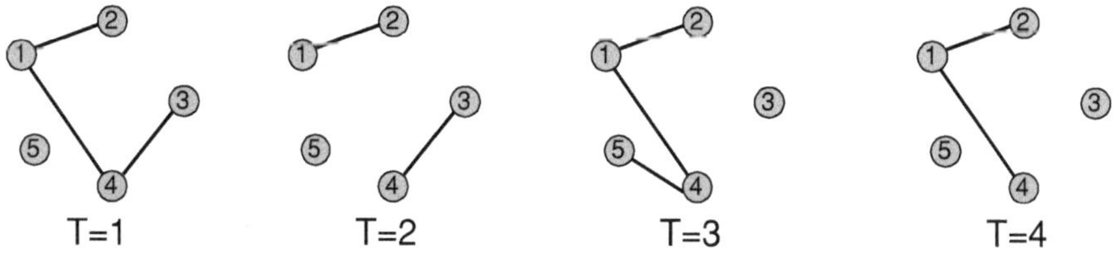

Fig. 4.5 Schematic illustration of the difference between considering observed ties or open ties in the definition of tie persistence. Although tie $1 \leftrightarrow 4$ is not observed at time $T = 2$, since it does not show any activity, it is however an open connection, since it is active both before and after $T = 2$

activity within $T = 2$, it is observed in both $T = 1$ and $T = 3$. Suppose now that we are using panel $T = 2$ as observation period to infer tie persistence or decay. Accordingly to common approaches where the tie persistence is inferred on the basis of its activity, tie $1 \leftrightarrow 4$ would be considered decayed while, instead, it is still active.

The consequences at global scale of assessing tie persistence/decay by considering observed ties or open ties are significant and lead to different picture of the cohesion and durability of the network. In Fig. 4.6 we show the fraction of surviving ties for the mobile phone network during a period of $T = 86$ weeks. Specifically, we divide the whole time period T in panels of 1 week each and consider the set of ties observed in the first 4 weeks. We then measure, in each panel, the fraction of this initial set of both the observed (black circles) and open (red squares) ties.
Although both curves can be approximated by a linear function $N(t) \sim bt$ with $b \simeq -0.006$, we observe a significant difference between them, signaling that two definitions of tie persistence yield to a significantly different picture of the persistence tie rate within a network. Specifically, due to the heterogeneous tie activity in both the number of calls (Miritello et al. 2012; Onnela et al. 2007) and the inter-event time

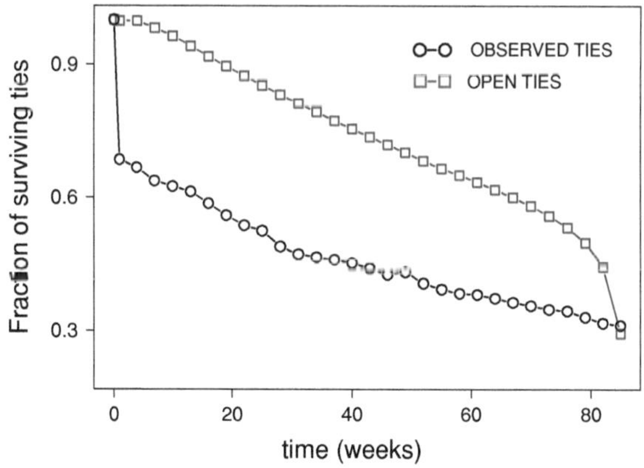

Fig. 4.6 Fraction of observed (*open circles*) and open (*open squares*) ties present in the first week of our observation period as a function of time over a period of 86 weeks. Both curves can be fitted by the $N(t) \sim bt$ with $b \simeq -0.006$. While persistence drops to 70 % after one month if ties are required to have activity at a given week, it is still around 70 % for one year if we consider open ties at that week

distribution (Barabási 2010; Miritello et al. 2011; Karsai et al. 2011), the number of observed ties across time is much smaller ($\sim 30 - 40\%$ less) than the actual number of persistent relationships. Note that the sharp drop in the number of observed ties in the first week is a clear manifestation of the bursty tie communication activity, which leads to the observation of almost half of the ties observed at the beginning.

In line with what observed in the previous chapter, this result is a further evidence of the fact that the common approach of inferring the existence of a social tie in a given time period by only looking at its activity leads to a misleading representation of the social network. Instead, a more realistic picture is obtained by considering open ties, which is possible only after separating tie dynamics from tie activity. However, as previously discussed, this is possible only when a long period of observation T is available, in such a way that the time windows before and after T can be used to assess with better precision the time instants of tie creation and decay (see Sect. 3.2.2). Since usually the available data sets are restricted to short time periods, the latter is not always achievable. Nevertheless, the method we propose in the next section may offer a solution to this problem. In fact, if on one hand it addresses the problem of how temporal patterns of human communication may help in the prediction of tie appearance/decay, at the same time it also represents a criterium to infer with good precision whether a tie observed in a given time window constitutes a new or old connection. As we will see, it can be thus used to separate tie dynamics from tie activity also when the observation period is relatively small (order of months).

4.4.2 Tie Prediction Based on Time-Dependent Features

From the data described in Appendix A, we consider the resulting phone communication network over a period of $T^* = 19$ months from February 2009 to August 2010. As mentioned above, we are interested in analyzing the role that the temporal patterns of communication tie observed within an observation window T in predicting whether they are more or less likely to have formed before T and, at the same time, to persist after T. To address this we follow the conventional approach discussed in Sect. 4.1. Specifically, once we have defined the prediction τ_1 and observation τ_2 periods, we develop a threshold criterium, similarly to what done by Hidalgo and Rodriguez-Sickert (2008), to assign to each of the link in τ_1 a persistent or decayed class. However, contrary to the traditional approaches, our criterium is based on temporal, instead than topological, features. Finally, to test the performance of our method, we compare the outcomes of our classifier with the empirical observation of the contact network in τ_2.
Thus, firstly, in a way analogous to what done in Chap. 3, we split the 19-months period into 3 subintervals [(Feb09–Jul09), (Aug09–Feb10), (Mar10–Aug10)] (see Fig. 4.7a). Again, to avoid the problem of inactivity/subscription/churn, which could alter the observed activity of a node/link, we only keep those nodes that at least have one communication event in each subinterval. We thus use the window T_2 in the middle (that for simplicity we denote T) as *prediction* time period τ_1 and the

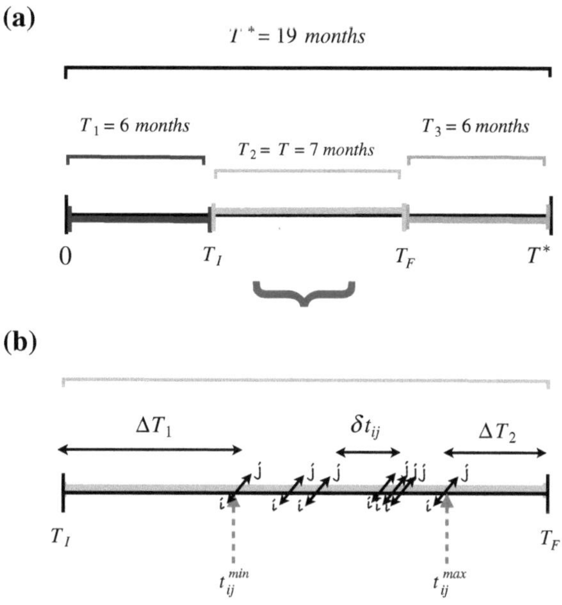

Fig. 4.7 **a** After dividing the whole observation time period of 19 months in three subintervals, we use the window T in the middle as predictive interval to measure tie features. The intervals before (T_1) and after (T_3) are then used to test our classifier and observe whether a tie exists from before or persists after T. **b** Tie activity in the predictive interval T. Slanting lines represent interaction events between i and j, which inter-event time is given by δt_{ij}

intervals before and after (respectively T_1 and T_3) as *observation* periods τ_2 to assess whether the ties present in τ_1 actually exist from before or persist after τ_1.

To show the idea of our method, let us consider the communication pattern of tie $i \leftrightarrow j$ in Fig. 4.7, where δt_{ij} are the inter-event time series which average is given by $\overline{\delta t}_{ij}$. We denote t_{ij}^{min} and t_{ij}^{max} the time stamps at which, respectively, the first and the last $i \leftrightarrow j$ communication are observed within the observation window T. We then assume that a tie exists from before T if the time until the first communication $\Delta T_1 = t_{ij}^{min} - T_I$ is small enough with respect to the average inter-event time $\overline{\delta t}_{ij}$. Analogously, if the time from the last communication $\Delta T_2 = T_F - t_{ij}^{max}$ is small enough with respect to $\overline{\delta t}_{ij}$, we assume that the tie will persist after T. In order for the average $\overline{\delta t}_{ij}$ to be significant, we only consider ties which have at least five communication events in the observation window T.

As mentioned in Sect. 2.4, the activity pattern of communication ties is high heterogeneous, indicating the existence of ties with different levels of activity, e.g. different number of calls w_{ij} in a given period T or, equivalently, different $\overline{\delta t}_{ij} = w_{ij}/T$. For this reason, one may think that the average $\overline{\delta t}_{ij}$ is not the more representative quantity for a given tie. The result in Fig. 4.8 justifies, however, our choice. Specifically, it shows the scaled probability density function $P(\delta t_{ij}/\overline{\delta t}_{ij})$ of the inter-event times for groups of ties with different average inter-event times $\overline{\delta t}_{ij}$. The bursty character of the tie activity is clearly demonstrated by the long tail of the distributions. However, regardless to the value of $\overline{\delta t}_{ij}$, all distributions collapse into a single curve. In line with previous results for mobile phone and e-mail communication (Candia et al. 2008; Goh and Barabási 2008; Karsai et al. 2011), this

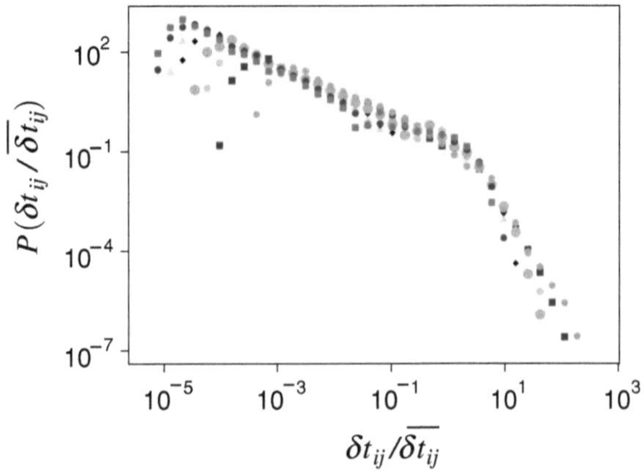

Fig. 4.8 Scaled inter-event time distribution for the tie activity. Different symbols refer to the pdf for groups of ties with different values of the average inter-event time $\overline{\delta t_{ij}}$, where each curve has been rescaled by the value of $\overline{\delta t_{ij}}$ of the correspondent bin

behavior indicates that the inter-event distribution is a universal function following the expression $P(\delta t_{ij}) = \left(1/\overline{\delta t_{ij}}\right) \mathcal{F}\left(\delta t_{ij}/\overline{\delta t_{ij}}\right)$, where $\mathcal{F}(x)$ is independent of the average activity.

4.4.3 Measuring the Performance of the Model

The problem of determining whether ties observed within a given window T exist from before (and/or persist after) T is now reduced to investigate the performance of our classifier $\Delta T_1/\overline{\delta t_{ij}}$ ($\Delta T_2/\overline{\delta t_{ij}}$) by comparing the outcomes with observations. For the latter task, we adopt a threshold-based model: to predict the existence before T we consider a discrimination threshold θ then assign to each tie the persistent class if $\Delta T_1/\overline{\delta t_{ij}} < \theta$ and the decayed class otherwise. Analogously, to predict the persistence of ties after T, we assign the persistence class if $\Delta T_2/\overline{\delta t_{ij}} < \theta$, the decayed otherwise. We then optimize the model by analyzing the performance of our classifier for different values of θ. After analyzing different values of θ, we found that interesting results are observed for $\theta \in [0, 100]$. This is the reason why in the following we show the results of our classifier only for this range of θ.

To this end, we validate the class assigned by our classifier with the empirical outcomes given by the observation of the network in the time windows T_1 and T_3. The prediction, that can be *positive* (the tie persists) or *negative* (the tie does not persist), will be considered *true* if correct and *false* otherwise. We then measure the rates of true positive (TP), true negatives (TN), false positives (FP) and false negative (FN). Specifically, in our case, for a given θ, a tie ij will contribute to the positives if $\Delta T_1/\overline{\delta t_{ij}} < \theta$ and to the negatives otherwise. After the validation, a positive-predicted tie can be true if that connection is actually observed and false otherwise. The same applies for the negatives. In general, a good classification model should be able to correctly capture not only the observations (TP) but also the non-observations

(TN). To analyze the performance of our classifier we measure the *precision* p (or positive predictive value) and the *recall* r (or true positive rate). While p gives the number of correct results divided by the number of all returned results $p = TP/(TP+FP)$, the recall r measures the number of correct results divided by the number of results that should have been returned $r = TP/(TP+FN)$. Precision and recall measure two competing principles, since a classification model could predict all ties as persistent thus achieving a very high recall which would result, however, in a very low precision. For this reason, usually, precision and recall scores are not discussed in isolation but combined into the $F_1\ score$, which captures the trade-off between them.

The $F_1 score$ is defined as the harmonic average between p and r and reaches its best value at 1 and worst at 0:

$$F_1\ score = \frac{2pr}{p+r} \tag{4.1}$$

Our results for the $F_1\ score$ in the range of $\theta \in [0, 100]$ are shown in Fig. 4.9. Red and blue curves refer respectively to the prediction of ties before and after T, which in principle do not have to give the best performance for the same value of θ. In the same figure we also show the *Mean Absolute Error* (MAE), that measures how close the predictions are to the outcomes. The MAE is defined as the average of the absolute errors e_i:

$$MAE = \frac{1}{n}\sum_{i=1}^{n} e_i = \frac{1}{n}\sum_{i=1}^{n} |f_i - y_i|, \tag{4.2}$$

where f_i is the prediction, y_i the true value and n the total number of ties.

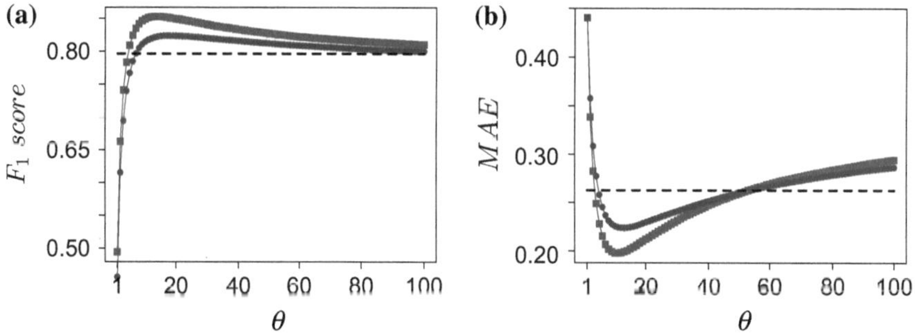

Fig. 4.9 Performance of the classifier $\Delta T_1/\overline{\delta t_{ij}}$ (*red curves*) and $\Delta T_3/\overline{\delta t_{ij}}$ (*blue curves*) in the prediction of ties, respectively, before and after the observation window T. In both cases the model performs particularly well for all the values of θ under consideration and, in particular, for $\theta \in [12, 16]$ where a maximum value of $F_1 \simeq 0.85$ ($F_1 \simeq 0.83$) **a** and a minimum error $MAE \simeq 0.19$ ($MAE \simeq 0.22$) **b** are respectively observed. *Dashed lines* show a comparison with the best performance obtained by considering a logistic regression model with predictors also the number of calls and the topological overlap. In the latter case a lower performance is obtained ($F_1 \simeq 0.80$ and $MAE \simeq 0.26$)

Table 4.1 Performance of the classifier to predict the ties appearance before and after T

Prediction after T	$\theta = 12$	$\theta = 13$	$\theta = 14$	$\theta = 15$	$\theta = 16$
Accuracy	0.806	0.807	0.808	0.807	0.807
Precision	0.807	0.803	0.799	0.795	0.788
Recall	0.848	0.856	0.864	0.870	0.880
F_1-score	0.827	0.829	0.830	0.831	0.831
MAE	0.224	0.223	0.222	0.222	0.223
Prediction before T					
Accuracy	0.801	0.801	0.799	0.797	0.795
Precision	0.812	0.807	0.801	0.796	0.792
Recall	0.895	0.904	0.910	0.916	0.921
F_1-score	0.852	0.852	0.852	0.852	0.852
MAE	0.198	0.199	0.200	0.202	0.204

Standard performance metrics for values of the discrimination threshold in the interval of $\theta \in [12, 16]$ where the model performs particularly well

It is particularly highlighting the observation of a maximum (minimum) in the F_1 $score$ (MAE), which indicates the existence of a particular value of the threshold that optimize the criterium. At the same time, we also observe a minimum in the absolute error. Specifically, for both problems of tie prediction before and after T, the model performs reasonably well for values of $\theta \in [12, 16]$, in which interval we found a maximum value of $F_1 \simeq 0.82$ ($F_1 \simeq 0.81$) and a minimum value of the error $MAE \simeq 0.19$ ($MAE \simeq 0.22$). The values obtained for the standard performance metrics in that range of θ are summarized in Table 4.1. Besides the quantities already defined, we also report the value of the *accuracy* of the classifier, which represents the proportion of outcomes classified correctly by the model and corresponds to the proportion of true results in the whole population.

As discussed in Sect. 4.2.1, most of the studies on the link prediction problem in social networks consider topological tie features, instead than time-dependent variables (Aiello et al. 2010; Akoglu and Dalvi 2010; Liben-Nowell and Kleinberg 2007; Lü and Zhou 2010; Hidalgo and Rodriguez-Sickert 2008; Raeder et al. 2011). For this reason, in Fig. 4.9 we also compare the performance of our model with the best performance obtained by applying a logistic regression model that takes as predictive variables $\Delta T/\overline{\delta t}_i j$, the number of communication events w_{ij} and the topological overlap O_{ij} (represented by dashed lines in Fig. 4.9). Among other variables, we have chosen w_{ij} and O_{ij} because these two quantities, together with tie reciprocity (that in our case is implied for any tie), have been found to have a particularly high predictive power (Akoglu and Dalvi 2010; Hidalgo and Rodriguez-Sickert 2008; Raeder et al. 2011). Even if a detailed analysis is out of the scope of this study, the results show that our simple criterium which takes into account the activity pattern of tie communication, gives better results than more complex models, typically used for evaluating the effectiveness of tie features in predicting tie appearance and decay (Raeder et al. 2011).

4.5 Discussion

In social networks, the link prediction problem is typically reduced to the problem of identifying the most important dimensions that characterize a tie in a given time window. This information is then used to build a predictive model to infer tie persistence or decay in a near future. The majority of studies in this area concentrate on characterizing *topological* properties of ties that lead to a satisfactory classification. It is well know, for example, that reciprocation, a high volume of communication or many common neighbors between two individuals decrease the chances for a relation to decay.

However, what has been ignored in most of these studies are the *temporal* properties of tie communication such as tie lifetime or the frequency of interaction. As well as static quantities, also temporal properties of tie communication must reflect the strength and the stability of the relationship. To shed light on this we first introduced and analyzed some time-dependent tie features that, contrary to static quantities such as the volume of communication, also capture the frequency and the homogeneity (heterogeneity) of communication. Specifically, we have studied the *temporal stability*, which measures tie lifetime, the *coefficient of variation* that gathers information about the burstiness of tie communication and the *average inter-event time* of interaction events. Although these concepts are not new, they have not previously investigated to characterize the strength of a social tie nor to predict future time appearance/decay. We measured these time features for all the communication ties of the phone call network described in Appendix A and analyzed their correlations with the standard topological features. Our results show that the current definition of social ties can be largely improved by taking into account not only *how much* two individuals communicate and *how many* contacts they share, but also *when* and *how* they interact.

Then, we investigated how temporal properties of communication ties can be used to predict which connections are more likely to be created or destroyed in the future. To address this, we first showed that time heterogeneity in human activity has important consequences on the definition of whether a social relationship is actually decayed or not. Specifically, we show that a measure of tie persistence based only on the observation of tie activity in a given period of time leads to a substantial underestimation of the number of persistent ties, since long periods of inactivity (bursts) can be confused with tie decay. This analysis is in line with what we have stressed in Chap. 3 and it constitutes a further evidence of the importance to separate tie activation/deactivation from their activity.

Finally, we proposed and studied a very simple method for tie prediction which allows to infer with high precision whether a tie is new/old by only looking at its communication activity. The idea of the model is the following: we assume that a tie is more likely to decay in the future if the time since the last communication observed within the observation period exceeds its average inter-event time. This idea is supported by the finding that the distribution of tie inter-event time $P(\delta t_{ij})$ can be described only in terms of the average value $\overline{\delta t}_{ij}$. We showed that this simple criterium performs better than more complex models that only include topological tie features, such as the volume of communication and the common neighbors.

The contribution of our results is twofold. On one hand, we presented a way to characterize tie strength and stability by means of simple measures that take into account also the temporal pattern of interaction. These features allow to distinguish between ties characterized by the same volume of communication, but with very different temporal patterns. This allowed to improve the current definition of a social tie and to characterize the underlying social relationship behind an observed activity pattern. On the other hand, we showed that current models of link prediction can be improved by including tie features that also take into account the temporal properties of tie communication. Actually, for its simplicity and generality, the simple criterium that we presented to predict tie appearance and decay can be applied to any tie classification problem involving temporal and dynamic networks. The implications of our results go beyond the link prediction problem, which constitutes a first step into the understanding of how to model dynamical social networks. It implies, in fact, the analysis of how (topological and temporal) properties of social relationships can be used to characterize, predict and model the evolution of the whole network.

References

Adamic L, Adar E (2003) Friends and neighbors on the web. Soc Netw 25(3):211–230

Aiello L, Barrat A, Cattuto C, Ruffo G, Schifanella R (2010). Link creation and profile alignment in the aNobii social network. In: Proceedings of the second IEEE international conference on social computing socialCom 2010, Minneapolis

Akoglu L, Dalvi B (2010) Structure, tie persistence and event detection in large phone and SMS networks. In: MLG '10 Proceedings of the eighth workshop on mining and learning with graphs

Aral S, Muchnik L, Sundararajan A (2009) Distinguishing influence-based contagion from homophily-driven diffusion in dynamic networks. Proc Natl Acad Sci USA 106(51):21544–21549

Barabási A-L (2010) Bursts: the hidden pattern behind everything we do. Dutton Books, New York

Burt R (1992) Structural holes: the social structure of competition. Harvard University Press, Cambridge

Burt R (2000) Decay functions. Soc Netw 22:1–28

Burt R (2001a) Attachment, decay, and social networks. J Organ Behav 22(6):619–643

Butts CT (2009) Revisiting the oundations of network analysis. Science 325:414–416

Candia J, González M, Wang P, Schoenharl T, Madey G, Barabási A-L (2008) Uncovering individual and collective human dynamics from mobile phone records. J Phys A Math Theor 41:224015

Crandall D., Cosley D., Huttenlocher D., Kleinberg J., and Suri S. (2008). Feed- back effects between similarity and social influence in online communities. In: KDD 08 Proceeding of the 14th ACM SIGKDD international conference on knowledge discovery and data mining New York, NY U. (ed), pp 160–168

Craven M, DiPasquo D, Freitag D, McCallum A, Mitchell T, Nigam K, Slattery S (2000) Learning to construct knowledge bases from the world wide web. Artif Intell 118:69–114

Eagle N, Pentland A, Lazer D (2009) Inferring friendship network structure by using mobile phone data. Proc Natl Acad Sci USA 106(36):15274–15278

Easley D, Kleinberg J (2010) Networks, crowds, and markets: reasoning about a highly connected world. Cambridge University Press, Cambridge

Getoor L, Diehl CP (2005) Link mining: a survey. ACM SIGKDD Explor Newsl 7:3–12
Gilbert E, Karahalios K (2009) Predicting tie strength with social media. In: CHI '09 Proceedings of the 27th international conference on human factors in, computing systems, pp 211–220
Glen J, Widom J (2002) SimRank: a measure of structural-context similarity. In: Proceedings of ACM SIGKDD international conference on knowledge discovery and data mining
Goh K-I, Barabási A-L (2008) Burstiness and memory in complex systems. Europhys Lett 81:48002
Gonçalves B, Perra N, Vespignani A (2011) Modeling users' activity on twitter networks: validation of dunbar's number. PLoS ONE 6(8):e22656
Granovetter M (1973) The strength of weak ties. Am J Sociol 78:1360–1380
Hidalgo C, Rodriguez-Sickert C (2008) The dynamics of a mobile phone network. Phys A 387:3017
Hill R, Dunbar R (2003) Social network size in humans. Hum Nat Interdisc Biosoc Perspect 14(1):53–72
Huang Z, Lin D (2009) The time series link prediction problem with applications in communication surveillance. Inf J Comput 21:286–303
Karsai, M, Kivelä, M, Pan R, Kaski K, Kertész J, Barabási A.-L. (2011) Small but slow world: how network topology and burstiness slow down spreading. Phys Rev E 83:025102(R)
Karsai M, Kaski K, Barabási A-L, Kertész J (2011) Universal features of correlated bursty behaviour. Sci Rep 2:397
Katz L (1953) A new status index derived from sociometric analysis. Psychometrika 18:39–43
Kossinets G, Watts DJ (2006) Empirical analysis of an evolving social network. Science 311:5757
Kossinets G, Watts DJ (2009) Origins of homophily in an evolving social network. Am J Sociol 115(2):405–450
Kovanen L, Saramäki J, Kaski K (2011) Reciprocity of mobile phone calls. Dyn Socio-Econ Sys 2:138–151
Lemaire V, Hue C, Bernier O (2009) Correlation exploration in a classification model. In: KDD '09 Proceedings of the 15th ACM SIGKDD international conference on knowledge discovery and data mining
Liben-Nowell D, Kleinberg J (2007) The link-prediction problem for social networks. In: CIKM '03: proceedings of the twelfth international conference on Information and knowledge management, New York, pp 556–559
Lü L, Zhou T (2009) Role of weak ties in link prediction of complex network. In: CNIKM '09 proceedings of the 1st ACM international workshop on Complex networks meet information and, knowledge management, pp 55–58
Lü L, Zhou T (2010) Link prediction in weighted networks: the role of weak ties. Europhys Lett 89:18001
McPherson J, Smith-Lovin L, Cook J (2001) Birds of a feather: homophily in social networks. Annu Rev Sociol 27:415–444
Miritello G, Lara R, Moro E (2013) Time allocation in social networks: correlation between social structure and human communication dynamics. In: Holme P, Saramäk, J (eds) Temporal networks. Springer, Heidelberg
Miritello G, Moro E, Lara R (2011) Dynamical strength of social ties in information spreading. Phys Rev E 83:045102(R)
Miritello G, Lara R, Cebrián M, Moro E (2012a) Limited communication capacity unveils strategies for human interaction. arXiv:1304.1979
Newman MEJ, Park J (2003) Why social networks are different from other types of networks. Phys Rev E 68:036122
Newman MEJ, Park Y (2004) The statistical mechanics of networks. Phys Rev E 70:066117
O'Madadhain J, Hutchins J, Smyth P (2005) Prediction and ranking algorithms for even-based network data. In: Proceedings of the SIGKDD explorations, vol 7
Onnela J-P, Saramäki J, Hyvönen J, Szabó Z, Lazer D, Kaski K, Kertész J, Barabási A-L (2007) Structure and tie strengths in mobile communication networks. Proc Natl Acad Sci USA 104:7332
Page L, Brin S, Motwani R, Winograd T(1999) The page rank citation ranking: bringing order to the web. Stanford Digital Library Technologies Project

Palla G, Barabási A-L, Tamás V (2007) Community dynamics in social networks. Noise Stochast Complex Syst Fin 6601:660106
Raeder T, Lizardo O, Chawla N, Hachen D (2011) Predictors of short-term decay of cell phone contacts in a large scale communication network. Soc Netw 33:245–257
Richardson M, Domingos P (2006) Markov logic networks. In: Machine learning
Roth M, Ben-David A, Deutscher D, Flysher G, Horn I, Leichtberg A, Leiser N, Matias Y, Merom R (2010) Suggesting friends using the implicit social graph. In: KDD '10 proceedings of the 16th ACM SIGKDD international conference on knowledge discovery and data mining, pp 233–242
Salton G, McGill MJ (1983) Introduction to model information retrieval. MuGraw-Hill, Auckland
Saramäki J, Leicht E, Lopez E, Roberts S, Reed-Tsochas F, Dunbar R (2012) The persistence of social signature in human communication. arXiv:1204.5602
Sarukkai RR (2000) Link prediction and path analysis using markov chains. In: Proceedings of the word wide web conference on computer networks machine learning
Song H, Cho T, Dave V, Zhang Y, Qiu L (2009) Scalable proximity estimation and link prediction in online social networks. In: IMC the internet measurement conference
Suitor J, Wellman B, Morgan D (1997) It's about time: how, why, and when networks change. Soc Netw 19:1–7
Tantipathananand C, Berger-Wolf T, Kempe D (2007) A framework for community identification in dynamic social networks. In: Proceeding KDD '07 proceedings of the 13th ACM SIGKDD international conference on knowledge discovery and data mining, pp 717–726
Vázquez A, Dobrin R, Sergi D, Eckmann J-P, Oltvai Z-N, Barabási A-L (2004) The topological relationship between the large-scale attributes and local interaction patterns of complex networks. Proc Natl Acad Sci USA 101:17940–17945
Wang C, Satuluri V, Parthasaraty S (2007) Local probabilistic models for link prediction. In: IEEE-ICDM international conference on data mining, vol 7
Wellman B, Wortley S (1990) Different strokes from different folks: community ties and social support. Am J Sociol 96:558–588
Yu K, Chu W, Tresp V, Xu Z (2007) Stochastic relational models for discriminative link predicion. In: Advances in neural information processing systems
Zhou T, Ren J, Medo M, Zhang Y-C (2007) Bipartite networks projection and personal recommendation. Phys Rev E 76:046115
Zhou T, Lü L, Zhang Y-C (2009) Predicting missing links via local information. Eur Phys J B 71:623–630

Chapter 5
Information Spreading on Communication Networks

If we can understand how behaviors spread...we could potentially promote behaviors like...condom use or tolerance.
—Sinan Aral

A quantitative analysis of human communication patterns is essential not only for a better understanding of human behavior, but also to explain the dynamics of many social, technological and economic phenomena. Examples include epidemics spreading (Anderson and May 1992; Bailey 1975), spread of news and opinions (Adar et al. 2004; Leskovec et al. 2009), diffusion of innovations (Díaz-Guilera et al. 2009; Guardiola et al. 2002; Rogers 1995), rumors or trends (Lazer et al. 2009; Pastor-Satorras and Vespignani 2001b). All these processes are shaped by the underlying structure of the network and by the temporal activity patterns of humans, since they depend on the way humans interact and share information (Barabási 2005; Iribarren and Moro 2009; Malmgren et al. 2008). A large understanding of spreading phenomena comes from implementing epidemiological models on empirical and synthetic social networks, by assuming that information among individuals propagates like a epidemic disease (Anderson and May 1992) This hypothesis implies homogeneity in both the network topology and in the timing of interaction events: each individual can interact to everybody else at any time instant. Due to the fact that they can be usually treated analytically, "epidemiological" approaches provide a simple way to characterize and model spreading phenomena. However, this constitutes the first approximation to real diffusion processes, where all nodes and ties are treated equally. In fact, as seen in the previous chapters, the topology of real social networks is highly heterogeneous: some individuals are much more connected than others, the flux of information that passes through each connection is unevenly distributed, social relationships are organized into community structures, etc. It is therefore logical to expect that the probability that an individual passes a piece of information to others individuals also depend on the network structure heterogeneities, e.g. to who each individual is connected to and how. In order to gain insight on the spreading properties of real phenomena, many efforts have been made in recent

G. Miritello, *Temporal Patterns of Communication in Social Networks*, Springer Theses, DOI: 10.1007/978-3-319-00110-4_5,

years to understand and quantify the impact of the complex topological patterns of the underlying network on spreading phenomena (Barrat et al. 2008; Barthélemy et al. 2004; Boguña and Pastor-Satorras 2002; Boguña et al. 2003; Danon et al. 2008; Saramäki and Kaski 2005; Pan et al. 2011). These studies allowed to take a step forward into the comprehension of the dynamics of many real spreading processes.

However, paradoxically, most of these studies neglect the *temporal patterns* of human communications discussed in Sect. 2.4: humans act in bursts or cascades of events (Barabási 2005; Isella et al. 2011; Rybski et al. 2009; Vázquez et al. 2007), most ties are not persistent (Hidalgo and Rodriguez-Sickert 2008; Kossinets and Watts 2006) and communications happen mostly in the form of group conversations (Eckmann et al. 2004; Isella et al. 2011; Zhao and Oliver 2010; Wu et al. 2010). However, since information transmission and human communication are concurrent, the temporal structure of communication must influence the properties of information spreading. Indeed, recent experiments of electronic recommendation forwarding (Iribarren and Moro 2009) and simulations of epidemic models on email and mobile databases (Karsai et al. 2011; Vázquez et al. 2007) found that the asymptotic speed of information spreading is controlled by the bursty nature of human communications that leads to a slowing down of the diffusion. Nevertheless, although the asymptotic speed is an important property of the propagation of information in social networks, there is still no general understanding of what temporal properties of human communication do influence spreading processes and how.

This is, indeed, what we address in this chapter. Without claiming to be exhaustive, we first give an overview of some of the basic models typically used to describe spreading processes on real networks and discuss the main results present in the literature on the role played by the network topology. Then, we present our main contribution to this topic which aims to quantify and model the impact of dynamical patterns of human communication in the process of information spreading in real social network. To this end, we simulate the SIR (Susceptible-Infected-Recovered) model on a large phone call data set over a long period of time. The large-scale and the fine-grained temporal resolution of the dataset us great advances in understanding how information may propagate from person to person, since it incorporates both the topological and the temporal properties of human interaction. Our results are twofold. First, we investigate how the bursty nature of human dynamics and the group conversations affect the speed and the reach of the information spreading. Specifically, we show that the interplay of these two effects is crucial for the reach of information: while the burstiness of human communication hinders the propagation of information through social ties, the correlations between communication events in neighbor ties (groups conversations) favor the formation of local information cascades. These two competing effects shape the spreading reach and the speed of information yielding to different possible behaviors depending on the temporal properties of the transmission process (Miritello et al. 2011).

We also investigate the effect of the real tie creation/removal in spreading phenomena. Together with the burstiness and group conversations, the latter dynamics constitutes one of the main inherent temporal inhomogeneities of human behavior, which has been usually neglected in the modeling of social networks and dynamical processes.

Indeed, as we have seen in Chap. 3, tie formation and decay is far from random and its dynamics crucially affect the characterization of fundamental quantities such as the social connectivity. Here we show that this dynamics also plays an important role in spreading phenomena. Specifically, we observe that the way in which people establish and destroy social relationships has a slowing-down effect in the way information spread. Taken together, these results suggest that temporal patterns of communication must be incorporated in the description and modeling of dynamical human-driven phenomena.

Second, we propose a way to map the dynamics of human interactions onto a static representation of the social network by means of the *transmissibility* of a tie, which represents the probability that a piece of information will be transmitted though that connection. Since the transmissibility incorporates both topological and temporal patterns of tie communication, we refer to this quantity as *dynamical strength* of ties, in contrast to the *static strength* given by the aggregated volume of communication and typically used to assess the weight of social ties. As we will see, this approach provides a quantitative description of information propagation by successfully predicting the percolation threshold of the spreading process in temporal networks (Miritello et al. 2011). More importantly, it also constitutes a step forward towards the description of social networks as dynamical objects, instead than static entities.

5.1 Modeling Information Spreading Phenomena

Since the 18th century mathematical models have been applied to real spreading phenomena in order to understand and control their mechanisms and effects. Most of the standard approaches came from epidemiological models since, for obvious reasons, a particular focus has been placed on the spread of infectious diseases (Anderson and May 1992; Bailey 1975; Murray 1993). In this study however we are more interested in modeling information (or rumor) spreading processes in which human communication interactions play an essential role, such as the spread of innovations, adoption of opinions and technologies, identification of social influence in marketing and more (Daley and Kendall 1964; Goffman and Newill 1964). Edges in such networks have different interpretations. While in disease propagation edges can be interpreted as *who infects whom*, in information spreading they represents *who gets information from whom* or *who influences whom*. On the other hand, when modeling information spreading phenomena that involve human interactions, in many cases it is desirable to spread the "epidemic" as fast and as efficiently as possible, rather than preventing the propagation as it happens for epidemic diseases or computer viruses. As in the case of many marketing campaigns or protocols for data dissemination, the problem consists of designing an epidemic algorithm in such a way that the dissemination of data or information from any node of a network reaches the largest number of remaining nodes as quickly possible.

In recent years information diffusion and social influence have been modeled for a wide range of empirical data, such as email (Liben-Nowell and Kleinberg 2008), online social networks (Aral and Walker 2012; Bakshy and Rosenn 2012) and mobile phone interaction (Karsai et al. 2011; Miritello et al. 2011; Onnela et al. 2007). Unfortunately, understanding the very mechanisms that drives information diffusion and influence is still a not fully understood issue. There are at least three fundamental reasons for this lack of understanding. First, observing individual transmission is typically very difficult due to the lack of data, which prevents to track the cascading processes taking place in the network. As well as epidemiologists that can observe when a person becomes ill without knowing who infected her or how much time was necessary for the infection to take hold, observing a diffusion process often reduces to noting only when nodes reproduces a piece of information, buy a product or subscribe to a service. Viral marketers, for example, can track when customers buy a product or subscribe to services, but typically they can not observe who or what influences the customers' decisions or when they passed the information to whom. In all these scenarios, therefore, the mechanism underlying the process is hidden and no information is provided about the source or the time it took to pass the information. This mechanism is however of outstanding interest and necessary for stopping infections or maximizing the propagation of a piece of news, idea or the sale of a product. Another factor that hinders the understanding of how (and why) information propagates between people is the presence of homophily in social networks, i.e. the tendency of individuals with similar characteristics to form relationships with one another (Kossinets and Watts 2009; McPherson et al. 2001). Due to the fact that humans tend to engage in similar activities as their contacts, it is often impossible to determine whether a particular interaction between two individuals, a common behavior or the acquisition of the same product can be attributed to influence/contagion or not. Indeed, it can just be a reflection of the similarity between the involved individuals (Aral et al. 2009; Manski 1993; Shalizi and Thomas 2011) or a consequence of their intense interaction, since individuals that interact often have a greater probability to influence one another (Bakshy and Rosenn 2012). Although some methods have been recently proposed to separate peer influence from homophily factors in observational data (Aral et al. 2009; Aral and Walker 2011), the question of how information diffuses through the network still remains.

The third challenge is related to the fact that diffusion processes are usually assumed to occur over a static and fixed contact network that does not change over time. This leads to the idea that, once the interaction network is revealed (who interacts with whom), the spreading process can happen at any time through any edge, with a probability which is fixed or depending on the (aggregated) strength of the tie (Onnela et al. 2007). As a consequence, the paths identified as potential channels over which information propagates, do not take into account temporal variations of the contact network nor the fact that interactions are time-ordered. Time, however, plays an essential role in the diffusion of diseases, information and influence over networks being all these processes usually driven by a causal ordering of interactions.

In addition, sometimes, the process of information diffusion itself can modify the structure of the network, a process which is known as coevolution of the social network (Holme and Newman 2006).

5.1.1 Uncorrelated and Static Networks

The standard theoretical approach to epidemic spreading is based on compartmental models, where individuals in the population are divided into different groups (Anderson and May 1992). Each group contains individuals at different states of the infection dynamics and individuals can move between some of the compartments as they change their state. One of the basics assumption of compartmental models is that the population is fully-mixed (homogeneous mixing hypothesis), meaning that the individuals with whom a susceptible individual has contact are chosen at random from the whole population. Therefore, the probability of contact between two individuals at different states depends on the densities of individuals in the different compartments.

The simplest formulation of compartmental models is the SI (Susceptible-Infected) model in which a population of N individuals is divided into two states: susceptible (S) and infected (I). Susceptible individuals become infected at a given transmission rate λ by interacting with an infected individual. According to this process, all the nodes will be eventually infected with a speed that only depends on the network structure and the epidemic dies once the whole population has been reached. In many real cases, an individual can catch the same disease more than once, i.e. tuberculosis or computer viruses (Pastor-Satorras and Vespignani 2001a,b). These cases are instead better described by the SIS (Susceptible-Infected-Susceptible) model: susceptible individuals can be infected and become again susceptible after a period of time in which they recover, thus being again exposed to the epidemic. This model is a model of endemic diseases since the disease can circulate around the population and persist indefinitely.

A richer (and often more realistic) dynamics is given by the SIR (Susceptible-Infected-Recovered) model. In this model each individual can be in one of three possible states, susceptible (denoted by S), infected (I), or recovered (R). Individuals still obey the SI dynamics, but once they are infected with transmission rate λ, they are now allowed to recover and acquire immunity (or die) at some recovery rate μ. Under the fully mixed hypothesis, the SIR model reduces to a system of coupled nonlinear differential equations (Bailey 1975):

$$\frac{ds}{dt} = -\lambda is; \qquad \frac{di}{dt} = -\mu i + \lambda is \qquad \frac{dr}{dt} = \mu i, \tag{5.1}$$

where $s(t)$, $i(t)$ and $r(t)$ are the fraction of population in each of the three states at time t and one of the equations is redundant since $s(t) + i(t) + r(t) = 1$ necessarily at all times. Equation (5.1) can be interpreted as follows: infected individuals decay into the removed class at a rate μ, while susceptible individual become infected at a rate proportional to both the densities of infected and susceptible individuals.

One of the most significant predictions of Eq. (5.1) is the existence of a epidemic threshold λ_c, which gives information on whether the initial individual (or seed) of infection gives rise or not to an infinite epidemic that involves a macroscopic part of the population (Murray 1993). If $\lambda > \lambda_c$ the process percolate through the whole population. On the other hand, when $\lambda < \lambda_c$, the number of infected individuals is infinitesimally small in the limit of very large populations ($N \to \infty$), thus the disease involve only a local part of the network and the process may die out. The epidemic threshold λ_c is related to the *basic reproductive number* R_0, which is another important quantity in epidemiological modeling. R_0 is defined as the expected number of secondary infections produced when one infected individual is introduced into a population where everyone is susceptible (Anderson and May 1992). If $R_0 > 1$ the infection reaches a significant fraction of the population (*tipping point*), while if $R_0 < 1$ the propagation dies quickly. For heterogeneous networks, the basic reproductive number R_0 is equal to $R_0 = \lambda\mu^{-1}\langle k^2\rangle/\langle k\rangle$ (Anderson and May 1992). Since in the real world an entire population is rarely totally susceptible, one usually measures the *secondary reproductive number* R_1, that is the actual average number of secondary infections per primary case observed and it is typically smaller than R_0. As R_0, the quantity R_1 gives information about the percolation transition in the SIR process (which happens at $R_1 = 1$ (Newman 2002)), but also about the speed of diffusion (which is proportional to R_1 (Barthélemy et al. 2004)) and the size of the cascades (which is a growing function of R_1 (Newman 2002)).

The SIR model can be solved exactly on a wide variety of networks. Nevertheless, it constitutes the first approximation to real networks which (i) neglects the possibility of correlation within the network structure and (ii) assumes homogeneity in the timing of events, modeled by a homogeneous processes. One in fact assumes that all individuals have approximately the same number of contacts in the same time and that all contacts transmit the disease and recover with the same probability. Indeed, since λ and μ are two fixed constants, no heterogeneity is allowed neither in the transmission rate nor in the duration of the infection or recovery time $T \sim 1/\mu$. In real life, however, all these assumptions fail. For this reason, generalizations of the SIR model in which times and probabilities are nonuniform and correlated have been proposed (Newman 2002).

5.1.2 The Role of Topological Properties

The existence and the value of the epidemic threshold λ_c strictly depends on the topology of the network. Specifically, for uncorrelated heterogeneous graphs with a generic degree distribution $P(k)$ and a finite average connectivity $\langle k\rangle$ one gets that λ_c is inversely proportional to the connectivity fluctuations (Moreno et al. 2002):

$$\lambda_c = \frac{\langle k \rangle}{\langle k^2 \rangle}, \tag{5.2}$$

below which the epidemic incidence is null, and above which it attains a finite value. For networks with $\langle k^2 \rangle < \infty$ the threshold has therefore a finite value. However, for heterogeneous networks (as for example scale free networks with $2 < \gamma \leq 3$) Eq. (5.2), together with the fact that $\langle k^2 \rangle \rightarrow \infty$, leads to a vanishing epidemic threshold.

The divergence of $\langle k^2 \rangle$ also yields to a basic reproductive number R_0 that always exceeds unity. As a consequence, epidemics always have a finite probability to survive indefinitely, whatever the spreading rate λ is (Pastor-Satorras and Vespignani 2001a,b; Moreno et al. 2003). This is a very relevant result, signaling that the high heterogeneity of scale free networks makes them extremely weak with respect to infections, a result that has several implications in human and computer virus epidemiology (Lloyd and May 2001). Network degree fluctuations also play a crucial role in the time scale τ that governs the reach of infection. Specifically, it has been found that the number of infected individuals $i(t)$ grows in time as $i(t) \simeq i_0 \exp t/\tau$, where i_0 is the initial density of infected individuals and τ behaves like $\tau \sim \langle k \rangle / \langle k^2 \rangle$. This implies that in networks with a scale-free structure epidemics spreads almost instantaneously. This feature is associated to the fact that the spreading follows a precise hierarchical dynamics and once the highly connected hubs are reached, the infection propagates until the smaller degree nodes and pervades the entire network (Barthélemy et al. 2004). The divergence of $\langle k^2 \rangle$ for scale free networks with $2 < \gamma \leq 3$ is a sufficient condition for a null epidemic threshold, since the rise of epidemic incidence is instantaneous (Boguña and Pastor-Satorras 2002; Eguiluz and Klemm 2002). However, for a general network in which the degrees of vertices are correlated, the above picture is not correct (Boguña et al. 2003). In the case in which degree correlations are explicitly controlled by the conditional probability $P(k'|k)$ that a node of degree k is connected to a node with degree k' (see 2.1.1), the epidemic threshold for both the SIS and the SIR models is inversely proportional to the largest eigenvalue of the connectivity matrices $C_{kk'} = kP(k'|k)$ and $C_{kk'} - k(k'-1)P(k'|k)$, respectively. In all these cases the epidemic threshold is determined by $C_{kk'}$ and not by the connectivity distribution $P(k)$ and, depending on the nature of the correlations, a positive value of λ_c can be obtained (Boguña and Pastor-Satorras 2002).

Taken together, these results show that the network structure can strongly affect spreading dynamics. This is especially important in all those processes where biological or electronic viruses, rumors or piece of information are transmitted through the edges, whose correlations reflect the non trivial structure of real networks.

For all such processes not only the degree distribution and correlations affect the reach and the speed of the spreading, but also features as the short path length (Watts and Strogatz 1998), node distance (Kitsak et al. 2010), community structure or correlations between network topology and tie strengths (Grabowicz et al. 2012; Onnela et al. 2007; Park et al. 2010).

5.1.3 The Impact of Non-poissonian Activity Patterns

The models described above, either including or not topological aspects of real networks, suppose a "*poissonization*" of the dynamical network. This assumption implies that: (i) communication can happen at any time, (ii) all interactions between agents take place uniformly in time, (iii) the interaction process is markovian or memoryless, (iv) there is no causality in the interaction events. In real life, however, none of these assumption holds. As we have seen in Sect. 2.4.3, many real networks show temporal inhomogeneities and correlations that, together with the fact that spreading processes happen accordingly to the time ordering of events, have important effects on the dynamics of spreading on real networks. Consequently, the empirical dynamics of spreading processes significantly differs from poissonian expectations. Specifically, it has been found that the bursty activity pattern of human communication makes real viral cascades last much longer than expected by assuming a Poisson-like dynamics (Iribarren and Moro 2009; Karsai et al. 2011; Vázquez et al. 2007). In fact, as we have seen, the probability of one node to interact with another within a time interval τ can be approximated by the waiting time equation in renewal processes (Eq. 2.12). Assuming that users may become infected at random times, the average τ is given by the solution of Eq. (2.12): $\overline{\tau} = \overline{\delta t}/2\,(1 + \sigma^2_{\delta t}/\overline{\delta t}^2)$, where δt is the inter-event time (Breuer and Baum 2005). Therefore, while in a Poissonian process $\sigma^2_{\delta t} = \overline{\delta t}^2$ thus $\overline{\tau} \simeq \overline{\delta t}/2$, in many real systems $\sigma^2_{\delta t} > \overline{\delta t}^2$ due to the burstiness of δt, which makes the response time much bigger. This explains why in real spreading processes information travels slower than expected, leading to the fail of the traditional epidemic models. As shown in Fig. 5.1, a better model for the observed slow dynamics is given by the non-markovian Bellman-Harris branching model (Harris 2002) by using a lognormal (rather than an exponential) distribution of responses times that incorporates the high variability of human behavior (Iribarren and Moro 2009). This model has been proved to fit perfectly the experiments of controlled viral marketing campaigns done in different countries (Iribarren and Moro 2011).

5.2 Information Spreading in Communication Networks

As we have seen in Sect. 2.3, in most studies, real temporal activity is aggregated over time giving a static snapshot of the social interaction where ties are described by static strengths which do not include information about the temporal aspects of how humans interact. Temporal and topological aspects are therefore disentangled in the analysis. In this Section we merge both aspects in the case of information diffusion by adopting a functional definition of the social ties using the well-known map between dynamical epidemic models and static percolation (Newman 2002). The network is still described by a static graph, but the interaction strength between individuals now incorporates the causal and temporal patterns of their communications and

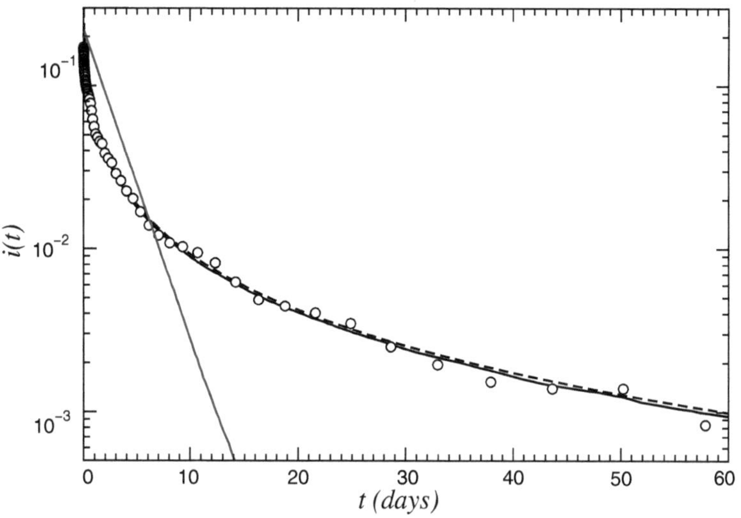

Fig. 5.1 Average fraction of participant as a function of time in a e-mail viral marketing campaign (*open circles*) compared with the prediction of the Bellman-Harris branching model with $P(t)$ the log-normal distribution (*black line*) and with $P(t)$ exponential with the same mean. Adapted from "Impact of human activity patterns on the dynamics of information diffusion" (Iribarren and Moro 2009)

not only the intensity (Onnela et al. 2007). This approach not only captures the temporal inhomogeneities due to the non-poissonian nature of human interactions, but also the topological aspects of the networks, their correlations and variations in time. We show that this procedure not only explains the qualitative behavior of the dynamics of information diffusion, but also successfully predicts the percolation threshold for the SIR model on empirical mobile telephone call records ad allows us to identify the relevant aspects of human communication in spreading processes (Miritello et al. 2011).

5.2.1 Characterizing Human Communication Patterns

By using the data set described in Appendix A, we firstly investigate the communication temporal patterns that might affect information diffusion. Spreading from user i to user j ($i \to j$) happens at the *relay time* intervals τ_{ij}, i.e. the time interval it takes to i to pass on to j an information he/she got from any another person $* \to i$, where $j \neq *$ (see Fig. 5.2). Information spreading is thus determined by the interplay between τ_{ij} and the intrinsic timescale of the infection process. As we have seen in Sect. 2.4.3, τ_{ij} depends on the correlated and causal way in which group conversations happen, since it depends on the inter-event intervals δt_{ij} in the $i \to j$ communication but *also* on the possible temporal correlation with the $* \to j$ events

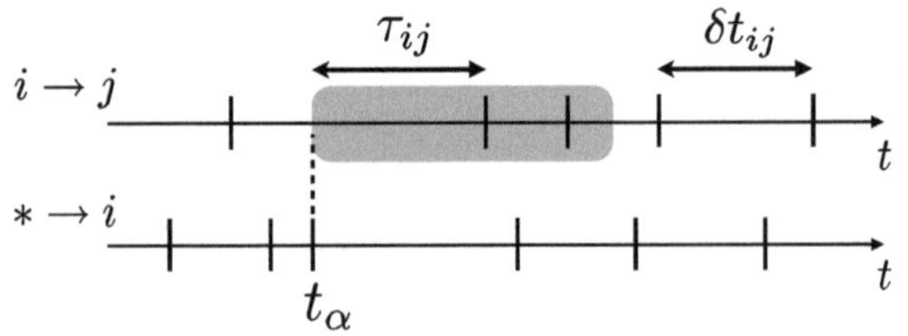

Fig. 5.2 Schematic view of communications events around individual i: each vertical segment indicates an event between $i \rightarrow j$ (*top*) and $* \rightarrow i$ (*bottom*). At each t_α in the $* \rightarrow i$ time series, τ_{ij} is the time elapsed to the next $i \rightarrow j$ event, which is different from the inter-event time δt_{ij} in the $i \rightarrow j$ time series. The red shaded area represents the recover time window T_i after t_α

(Newman 2002; Kenan and Robins 2007). By ignoring this correlation the probability distribution function (or pdf) for τ_{ij} can be approximated by the waiting-time density for δt_{ij} given by Eq. (2.12) that we rewrite here for convenience:

$$P(\tau_{ij}) = \frac{1}{\overline{\delta t_{ij}}} \int_{\tau_{ij}}^{\infty} P(\delta t_{ij})\, d\delta t_{ij}, \tag{5.3}$$

where $\overline{\delta t_{ij}}$ is the average inter-event time. In this approximation, the dynamics of the transmission process only depends on the dyadic $i \rightarrow j$ sequence of communication events and in particular, the possible heavy-tail properties of $P(\delta t_{ij})$ are directly inherited by $P(\tau_{ij})$. Figure 5.3 shows our (rescaled) results for $P(\delta t_{ij})$ and $P(\tau_{ij})$. For comparison, we also show the results obtained when *i)* the time-stamps of the $* \rightarrow i$ events are randomly selected from the complete CDR, thus destroying any possible temporal correlation with $i \rightarrow j$ and effectively mimicking Eq. (5.3) and *ii)* when the whole CDR time-stamps are shuffled thus destroying both tie temporal patterns and correlation between ties. Both shufflings preserve the tie intensity w_{ij}, i.e. the number of calls and their duration and also the circadian rhythms of human communication (Karsai et al. 2011). The result for $P(\delta t_{ij})$ shows that small and large inter-event times are more probable for the real series than for the shuffled ones, where the distribution is almost exponential as in a Poissonian process, apart from a small deviation at small times due to the circadian rhythms. This behavior is a reflection of the bursty pattern of activity in human interactions discussed in Sect. 2.4. However, here we see that it also happens at the level of the interaction between two individuals, confirming recent results in mobile (Karsai et al. 2011) and online communities (Rybski et al. 2009) dynamics.

As shown in the inset of Fig. 5.3, the distribution of τ_{ij} is also heavy-tailed but displays a larger number of short τ_{ij} compared to the shuffled one. The abundance of short τ_{ij} suggests that receiving an information ($* \rightarrow i$) triggers communication with other people ($i \rightarrow j$), a manifestation of group conversations (Zhao and Oliver 2010; Eckmann et al. 2004; Wu et al. 2010). While the fat-tail of $P(\tau_{ij})$ is accurately described by Eq. (5.3), and thus large transmission intervals τ_{ij} are mostly due to

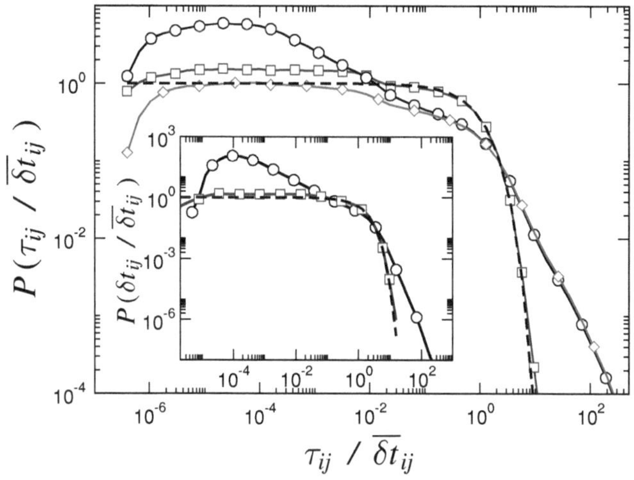

Fig. 5.3 Distribution of the relay time intervals τ_{ij} (main) and of the inter-event times δt_{ij} (inset) in the $i \rightarrow j$ tie rescaled by $\overline{\delta t_{ij}}$. The *black circles* correspond to the real data, while the *red squares* is the overall-shuffled result. *Blue diamonds* correspond to the case in which only the $* \rightarrow i$ sequence is randomized. Only ties with $w_{ij} \geq 10$ are considered. In both graphs the *dashed line* correspond to the e^{-x} function

large inter-event communication times in the $i \rightarrow j$ tie, the behavior of $P(\tau_{ij})$ is not only due to the bursty patterns of δt_{ij}, but also to the temporal correlation between the $i \rightarrow j$ and the $* \rightarrow i$ events. In fact, if the correlation between the $i \rightarrow j$ and the $* \rightarrow i$ series is destroyed, the probability of short-time intervals decreases and approaches the Poissonian case (see Fig. 5.3).

5.2.2 SIR Model on Real Networks

Our results show that relay times depend on two main properties of human communication that compete with one another. While the bursty nature of human activity yields large transmission times hindering any possible infection, group conversations translate into an unexpected abundance of short relay times, favoring the probability of propagation. To investigate the effect of these two conflicting properties of human communication on information spreading, we simulate the epidemic SIR model (see Sect. 5.1.1) in our social network considering the real time sequence of communication events and compare them to the shuffled data (Cebrián et al. 2009; Zhao and Oliver 2010; Karsai et al. 2011). We start the model by infecting a node at a random instant and considering all other nodes as susceptible. In each call an infected node can infect a susceptible node with probability λ. Due to the synchronous nature of the phone communication, this happens regardless of who initiates the call, so an infected node i can infect a susceptible node j either if he calls or is called from j.

However, we obtain the same results by considering directionality in the calls, i.e. i can infect j only if he initiates the call. For computational reasons, we therefore consider this latter case. Nodes remain infected during a time T_i until they decay into the recovered state. For the sake of simplicity we simulate the simplest model in which the recovering time T_i is deterministic and homogeneous $T_i = T$ and set $T = 2$ days, although different and/or stochastic T_i can be studied within the same model. The spreading dynamics generates a viral cascade that grows until there are no more nodes in the infected state. We repeat the spreading process for 3×10^4 randomly chosen seeds. Note that our model includes the SI model simulations in (Karsai et al. 2011) where $\lambda = 1$ and $T = T_0$, with T_0 being the total duration of the dataset.

We first ensure the existence of a percolation transition (Newman 2003) by looking at the size of the largest cascade s_{max} (over all realizations) at each value of λ. As shown in Fig. 5.4 the transition is confirmed by the observed change in the behavior of s_{max} from small to large cascades at a given value of λ (tipping point).

The same behavior is observed for the shuffled-time data where the transition seems to happen almost at the same value of λ (around 0.1) although an accurate analysis of the percolation point was beyond the scope of our study. In contrast, there is a significant difference in the behavior of the asymptotic average size s_∞ between the real and the shuffled-time data for different regimes of λ: when λ is small, s_∞ is larger for the real data than for the shuffled one, while the opposite behavior is observed for large λ. This difference, which can be very large for moderate values of λ, shows the impact of the real time dynamics of communication in the reach of

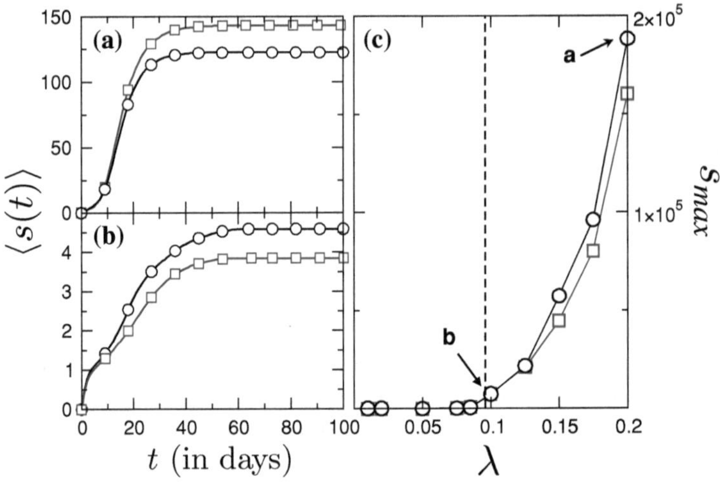

Fig. 5.4 Average size dynamics for a large (**a**) and a small (**b**) value of λ (*left*) and maximum size (*right*) of the infection outbreaks (over 10^4 realizations) for the real data (*black lines*) and shuffled data (*red lines*) for $T = 2$ days. The *dashed line* shows the critical point estimation of the percolation transition given by $R_1[\lambda, T] = 1$ with R_1 calculated using Eq. (5.18)

information in society. Specifically, if information propagates easily (large λ), thc average extent in temporal social networks is narrower than the one expected when a Poissonian dynamics is considered. In this sense, temporal patterns make social networks bigger than expected at large scales. However, in most real situations λ is very small (Iribarren and Moro 2009) and in this case the observed behavior is the opposite: despite the low propagation, information cascades are larger in real data than in the Poisson case, which suggests that information spreading is more efficient at small (local) scales.

5.2.3 The Dynamical Strength of Social Ties

To understand the observed behavior, we follow the approach of (Newman 2002) by mapping the dynamical SIR model to a static edge percolation model where each tie is described by the *transmissibility* $\mathcal{T}_{ij}$, that represents the probability that the information is transmitted from i to j and is a function of λ and T. If user i becomes infected at time t_α and the number of communication events $i \to j$ in the interval $[t_\alpha, t_\alpha + T]$ is $n_{ij}(t_\alpha)$, then the transmissibility in that interval is (see Fig. 5.2)

$$\mathcal{T}_{ij} = 1 - (1 - \lambda)^{n_{ij}(t_\alpha)}. \tag{5.4}$$

User i may become infected at any $* \to i$ communication event. Assuming these events independent and equally probable, we can average $\mathcal{T}_{ij}$ over all the t_α events to get

$$\mathcal{T}_{ij}[\lambda, T] = \langle 1 - (1 - \lambda)^{n_{ij}(t_\alpha)} \rangle_\alpha. \tag{5.5}$$

If the number of $* \to i$ events is large enough we can use a probabilistic description of Eq. (5.5) in terms of the probability $P(n_{ij} = n; T)$ that the number of communication events between i and j in a given time interval T is n. Thus

$$\mathcal{T}_{ij}[\lambda, T] = \sum_{n=0}^{\infty} P(n_{ij} = n; T)[1 - (1 - \lambda)^n], \tag{5.6}$$

which in principle can be non symmetric ($\mathcal{T}_{ij} \neq \mathcal{T}_{ji}$). This quantity represents the real probability of infection from i to j and defines what we called the *dynamical strength* of the tie. Note that $\mathcal{T}_{ij}$ depends on the series of communication events between i and j, but also on the time series of calls received by i. In (Newman 2002) Newman studied the case in which both time series are given by independent Poisson processes in the whole observation interval $[0, T_0]$. Thus, $P(n_{ij} = n; T)$ is the Poisson distribution with rate $\rho_{ij} = n_{ij}T/T_0$, where n_{ij} is total number of calls from i to j in $[0, T_0]$, and so

$$\tilde{\mathcal{T}}_{ij}[\lambda, T] = 1 - e^{-\lambda\rho} = 1 - e^{-\lambda w_{ij} T/T_0}, \tag{5.7}$$

which shows the one-to-one relationship between the intensity n_{ij} and the transmissibility $\mathcal{T}_{ij}$ in the Poissonian case: the more intense the communication is, the larger the probability of infection. However, as we have seen in Fig. 5.3, the real $i \rightarrow j$ and $* \rightarrow i$ series are far from being independent and Poissonian. To proceed analytically, we approximate Eq. (5.5). For small values of λ we have $1-(1-\lambda)^n \simeq \lambda n$, while when $\lambda \simeq 1$ we get that $1-(1-\lambda)^n \simeq 1$ for $n > 0$. Thus, the transmissibility for the two regimes is given by:

$$\mathcal{T}_{ij}[\lambda, T] = \begin{cases} \lambda \langle n_{ij} \rangle_{t_\alpha} & \text{for } \lambda \ll 1 \\ 1 - P^0_{ij} & \text{for } \lambda \simeq 1 \end{cases} \tag{5.8}$$

where $P^0_{ij} = P(n_{ij} = 0; T)$ is the probability of no event in the time interval T. This approximation allows us to to estimate T_{ij} in a much simpler way, since it depends now only on variables that can be measured from the temporal activity. Specifically, P^0_{ij} can be estimated directly from Eq. (5.3) for each link $P^0_{ij} = \int_T^\infty P(\tau_{ij})\, d\tau_{ij}$, since it measures the probability to find a relay time bigger than T. Figure 5.5 shows the comparison of n_{ij} and P^0_{ij} (averaged over all ties) for different values of T for the real and shuffled-time data (denoted by tilde). On the one hand, due to the correlation between the $* \rightarrow i$ and $i \rightarrow j$ time series, the number of events in a tie following an incoming call is always larger for the real data than for the shuffled one. This is the reason why, for small λ, the average transmissibility (and thus the size of the epidemic

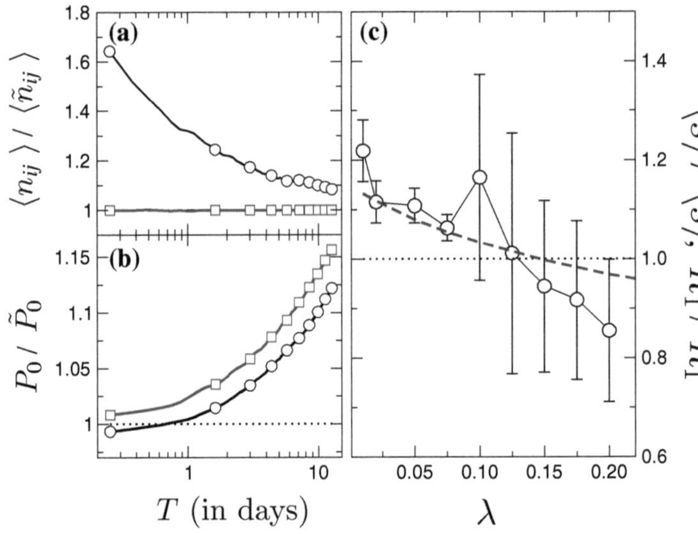

Fig. 5.5 Ratio of the number of events (**a**) and probability of no events (**b**) as a function of the recovery time T for the real (*black circles*) and shuffled $* \rightarrow i$ (*red squares*) data with respect to the overall-shuffled data. *Right panel* (**c**) shows the ratio of the average size of the outbreaks (*black circles*) and of R_1 calculated using Eq. (5.18) (*dashed blue line*)

cascades) is always higher in real communication patterns (Zhao and Oliver 2010). On the other hand, the bursty nature of the $i \to j$ communication makes the tail for the real $P(\tau_{ij})$ heavier than the exponential distribution found in the shuffled data. Thus, if T is large enough, P^0_{ij} is larger in the real-time than in the shuffled-time data and this is why we observe smaller cascades in that region. Note, however, that this does not apply for very small values of T ($T \lesssim 1$ days), where the causality between $* \to i$ and the $i \to j$ time series can make P_0 even smaller in the real case.

In Chap. 4 we have introduced the coefficient of variation cv_{ij} of the inter-event time distribution of a tie, as a measure of the burstiness of tie communication. Here we want to note that, for $\lambda \simeq 1$, the larger cv_{ij}, the larger the probability P^0_{ij}, thus the lower the transmissibility of the tie. The same applies for the stability ΔT_{ij}, which measures the lifetime of the tie (see Chap. 4): if the tie communication is concentrated in a short time period, then P^0_{ij} is large and, according to Eq. (5.8), the transmissibility is low. As a consequence, the shorter the lifetime of the tie, the smaller its transmissibility. These observations have a remarkable conclusion, since they indicate that the traditional description of a social network which neglect the temporal features of tie communication (see Sect. 2.3) typically overestimates the transmissibility power of social relationships. As expected, the temporal properties of tie communication also affect the total transmissibility of an individual, suggesting that the current vision of *social hub* as social spreader needs to be revised, something that we will discuss in Sect. 5.3.

To give a more quantitative analysis of the results in Fig. 5.4, we investigate the percolation process in a social network in which links have transmissibility $\mathcal{T}_{ij}$. Specifically, by following the approach by Newman (Newman 2002), we measure the secondary reproductive number R_1, i.e. the average number of secondary infections generated by each infectious individual. As mentioned in Sect. 5.1.1, R_1 gives information about the percolation transition in SIR processes: if $R_1 < 1$ the propagation dies quickly while if $R_1 > 1$ the infection reaches a significant fraction of the population (*tipping point*). The general idea is that, by definition, R_1 can be estimated if we know the following quantities:

1. the probability to randomly chose a link within the network;
2. the probability that the node i reached by the chosen link has degree k_i;
3. the excess-degree (or remaining degree) of the reached node i, i.e. the number of 'remaining' links (different from the ingoing one) that can be reached from i.

Let us first consider the simplest case in which the transmissibility is neglected, which corresponds to consider a fixed transmissibility equal for all ties ($\mathcal{T}_{ij} = \mathcal{T}$). In this case, the probability to randomly chose a link within the whole network is the same for all links. Once a link has been chosen, the probability $Q(k_i)$ to have an ending-node i with degree k_i is given by:

$$Q(k_i) = k_i/\langle k\rangle P(k_i), \tag{5.9}$$

where $P(k_i)$ is the network degree distribution and $(k_i - 1)$ the excess-degree of the chosen node. R_1 is thus straightforwardly obtained by:

$$R_1 = \int_{k_i} dk_i (k_i - 1) Q(k_i) \tag{5.10}$$

$$= \int_{k_i} dk_i (k_i - 1) \frac{k_i}{\langle k_i \rangle} P(k_i) \tag{5.11}$$

$$= \frac{\langle k_i^2 \rangle}{\langle k_i \rangle} \langle k_i \rangle. \tag{5.12}$$

In our case, however, each tie is characterized by a given transmissibility $\mathcal{T}_{ij}$, where in general $\mathcal{T}_{ij} \neq \mathcal{T}_{ik}$ for $j \neq k$. As a consequence, for each node we have a sequence of transmissibility $\{\mathcal{T}_{ij}\}_{j=1,\ldots,\,k_i}$, thus a distribution $P(\{\mathcal{T}_{ij}\})$. The procedure is the same than the previous case with the difference that now, instead of the degree k_i, each node is characterized by a collection of $\mathcal{T}_{ij}$. Therefore, after choosing a link within the whole network, we are now interested in the probability that the reached node i has a sequence of transmissibility equal to $\{\mathcal{T}_{ij}\}$. This probability is proportional to the number of links that can be reached from the chosen node and, in particular, it is given by:

$$Q(\{\mathcal{T}_{ij}\}) = \frac{\sum_j \mathcal{T}_{ij}}{\langle \sum_j \mathcal{T}_{ij} \rangle} P(\{\mathcal{T}_{ij}\}), \tag{5.13}$$

which is analogous to Eq. (5.9). Once reached a node with $\{\mathcal{T}_{ij}\}$, the excess-degree is instead given by $\sum_{j \neq l} \mathcal{T}_{ij}$, where l is the ingoing node of the chosen tie. There is however an important difference with the previous case. The probability of randomly chose a tie, in fact, is not the same for each tie as in the previous case and it depends, instead, on the transmissibility of the tie: the larger $\mathcal{T}_{li}$, the higher the probability of li to be chosen. Specifically, this probability is given by $\mathcal{T}_{il} / \sum_j \mathcal{T}_{ij}$. The excess-degree is therefore obtained by averaging this quantity over all ties l:

$$\sum_l \frac{\mathcal{T}_{ij}}{\sum_j \mathcal{T}_{ij}} \sum_{j \neq l} \mathcal{T}_{ij} = \frac{1}{\sum_j \mathcal{T}_{ij}} \sum_l \mathcal{T}_{il} \Big[\sum_j \mathcal{T}_{ij} - \mathcal{T}_{il} \Big] \tag{5.14}$$

$$= \frac{1}{\sum_j \mathcal{T}_{ij}} \Big[\Big(\sum_j \mathcal{T}_{ij} \Big)^2 - \sum_j \mathcal{T}_{ij}^2 \Big]. \tag{5.15}$$

By combining Eq. (5.13) with Eq. (5.15) and integrating over all the possible values of $\{\mathcal{T}_{ij}\}$, we have:

$$R_1 = \int_{\{\mathcal{T}_{ij}\}} d\mathcal{T}_{ij} \frac{\sum_j \mathcal{T}_{ij}}{\langle \sum_j \mathcal{T}_{ij} \rangle} P(\{\mathcal{T}_{ij}\}) \frac{1}{\sum_j \mathcal{T}_{ij}} \Big[\Big(\sum_j \mathcal{T}_{ij} \Big)^2 - \sum_j \mathcal{T}_{ij}^2 \Big] \tag{5.16}$$

$$= \frac{1}{\langle \sum_j \mathcal{T}_{ij} \rangle} \int_{\{\mathcal{T}_{ij}\}} d\mathcal{T}_{ij} P(\{\mathcal{T}_{ij}\}) \Big[\Big(\sum_j \mathcal{T}_{ij} \Big)^2 - \sum_j \mathcal{T}_{ij}^2 \Big], \tag{5.17}$$

which results in the approximation for R_1:

$$R_1[\lambda, T] = \frac{\langle (\sum_j \mathcal{T}_{ij})^2 \rangle_i - \langle \sum_j \mathcal{T}_{ij}^2 \rangle_i}{\langle \sum_j \mathcal{T}_{ij} \rangle_i}. \tag{5.18}$$

Note that in the homogeneous case, in which $\mathcal{T}_{ij} = \mathcal{T}$, we recover the common result for random networks $R_1 = \mathcal{T}(\langle k_i^2 \rangle / \langle k_i \rangle - 1)$ (Newman 2002). Figures 5.4 and 5.5 shows the accuracy of the approximations used to get Eq. (5.18) to predict the tipping point in the SIR process and the change in the average size of the cascades in the two regimes. This suggests that the dynamical strength of the ties $\mathcal{T}_{ij}$, defined in Eq. (5.5), can be effectively used to model real strength of human interactions in social networks.

5.3 Static Hubs and Dynamical Hubs

In previous chapters, we have seen that there are important correlations between the social connectivity of a given node and the properties of its social ties. We have shown, for example, that for large values of k_i, the time people dedicate on average to their social connections decreases with k_i (Sect. 3.1). In addition, also temporal properties such as tie lifetime and burstiness are distributed differently among neighbors for individuals with low and large social connectivity, as seen in Chap. 4. On the other hand, as mentioned in the previous section, these properties of tie communication make the transmissibility of a tie smaller than the one expected when only the number of interaction events is considered (Poissonian model). On the basis of these results, here we want to investigate whether the total transmissibility of an individual, which is related to the transmissibility of his connection by the $\mathcal{T}_i = \sum_j \mathcal{T}_{ij}$, correlates with the individual's social connectivity k_i. Note that, in contrast, in the Poissonian case and for small values of λ, we get that $\tilde{\mathcal{T}}_{ij} \simeq \lambda w_{ij} T / T_0$ and thus $\tilde{\mathcal{T}}_i \simeq \lambda\ s_i\ T/T_0$, where s_i is the node's strength introduced in Sect. 3.1. This means that the transmission power of a node is proportional to the strength of the node in this approximation.

In contrast, in the real case the transmissibility $\mathcal{T}_i = \sum_j \mathcal{T}_{ij}[\lambda, T]$ of a node depends both on the tie weights and dynamical patterns of communication. The relation between the Poissonian and the real transmissibility is shown in Fig. 5.6a. What we observe is that the total real transmissibility of an individual is smaller than the one obtained in the Possonian case, were the real time-stamps of tie communication have been shuffled. This result suggests that the effect of the burstiness of real tie communication is to make individuals less powerful to transmit information. Nevertheless, as shown in Fig. 5.6b, this effect is not equal for any node since the relative difference between $\mathcal{T}_i$ in the real and shuffled cases is larger for hubs or more connected people than for poorly connected people. This is a manifestation of the fact that hubs have shorter ties which, as discussed above, translate in a smaller

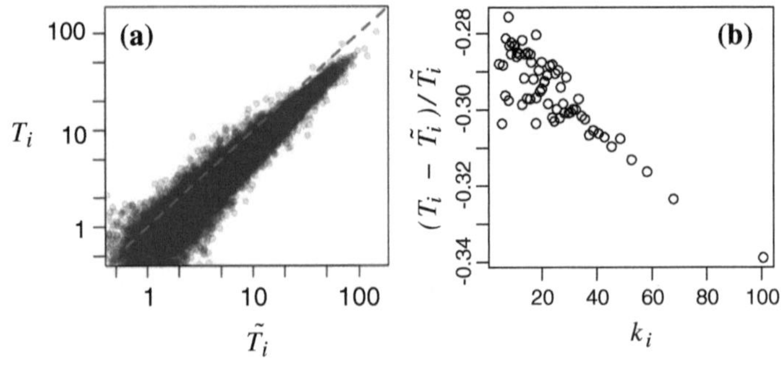

Fig. 5.6 **a** Comparison between the total transmissibility of individuals in the real T_i and shuffled $\tilde{T}_i$ call records for a random set of 10^5 users in our database. *Dashed line* is the $T_i = \tilde{T}_i$ line. **b** Conditional average of relative difference between the real and shuffled total transmissibility as a function of the social connectivity k_i

transmissibility. In addition, we want to stress that this relative difference increases with k_i, meaning that T_i does not grow linearly with k_i (or s_i) as, in contrast, $\tilde{T}_i$ does. Our findings show that neither k_i or s_i are good predictors of the local spreading power or influence of a node, specially for largely connected people or hubs. In this sense, although in general static hubs (people with large k_i or s_i) have also large dynamical transmissibility, the large variability shown in Fig. 5.6a implies that this correspondence is not always true (Miritello et al. 2013).

5.4 The Role of Ties Dynamics in Information Spreading

In Chap. 3 we have seen that the dynamics of tie creation and removal is highly articulated and happens at a time scale similar to the one that regulates the communication events. Here we analyze whether and how this dynamics affects the speed and the reach of information spreading. To address this, we employ a new reference model that allows us to separate the effect due to the burstiness from the one due to the ties creation and removal. In this new reference model we maintain the initial and the final time of each tie, such that the tie lifetime Δt_{ij} is preserved, but we shuffle the real time-stamps within Δt_{ij}, thus destroying the burstiness of human communication and conversations between people, i.e. possible correlations between events in different ties. Since the patterns of communication are now destroyed only within each tie, we refer to the latter reference model as *intra-tie shuffled data*, to distinguish it from the Poisson-like case where each time stamp is chosen from the whole CDR (*overall shuffled data*). Note that the intra-tie shuffling lies between the real-time case, where all the temporal features of human dynamics are preserved, and the Poisson-like data, where not only the burstiness of tie communication is destroyed, but also the tie creation/removal dynamics since all ties have the same probability to appear/disappear at any time. To investigate the effect of tie dynamics,

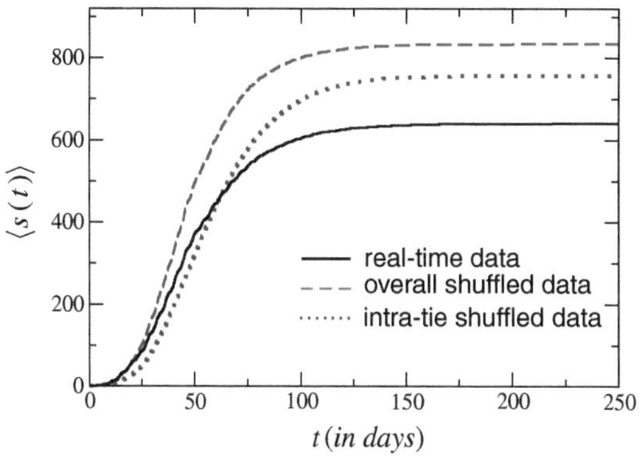

Fig. 5.7 Average fraction of infected nodes (over 10^4 realizations) as a function of time for $\lambda = 0.2$ and T=7 days obtained for real-time data (*solid curve*), shuffled-time data (*dashed curve*) and shuffled tie creation/removal data (*pointed curve*). The effect of tie creation/removal is to slow down the spreading dynamics

we then simulate the SIR model on the three time series by considering, once again, the simplest model in which the probability of infection λ is constant and the recovery time T deterministic and homogeneous. Figure 5.7 displays the average cascade size dynamics for $\lambda = 0.2$ and $T = 7$ days. This result allows us to gain insight into the effect of each temporal feature of human communication. In fact, the difference between the real and the intra-tie shuffled data is due to the bursty activity of ties, while the one between the Poisson-like and intra-tie shuffle is related to the tie formation/removal. As in the case of Fig. 5.4, for this value of λ (above the percolation point), information cascades are larger and the spreading faster for the Poissonian series than in the real one, due to the bursty activity patterns. However, when comparing the Poissonian case with the intra-tie shuffled one, we observe slower and smaller cascades in this latter case, indicating that the way in which individuals create and destroy relations has a slowing-down effect in the spreading dynamics.

5.5 Towards a Dynamical Model of Human Interactions

Our results indicate the necessity to incorporate temporal patterns of human activity in the description and modeling of human interactions. As discussed in Chap. 2, the static description of weighted networks implies a *poissonization* of human dynamics where the weight w_{ij} of a tie (aggregated volume of communication) can be seen as the rate of communication such that $P(i \rightarrow j)\,dt = \rho_{ij}\,dt$ and $\rho_{ij} = w_{ij}$. This approach neglects all those temporal patterns as the time correlations between events, the burstiness of interactions or the tie creation/removal that, as showed, have an important impact in the information speed and reach.

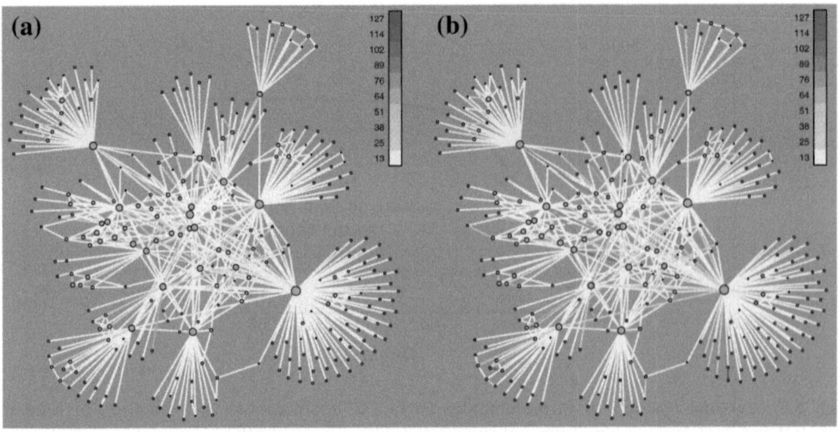

Fig. 5.8 Structure of the mobile phone network around a randomly chosen individual where the tie strength is taken as, **a** the effective weight $w^*_{ij}(\lambda, T) \sim \ln(1 - T_{ij}(\lambda, T))$ for $\lambda = 0.5$ and T = 15 days and **b** the total number of phone communication between i and j over the whole observation window (aggregated weight). Adapted from "Time allocation in social networks: correlation between social structure and human communication dynamics" (Miritello et al. 2013)

One simple way to account for the temporal properties of human communication and still give a static representation of the social network, is by using the transmissibility T_{ij} as a measure of tie strength, instead of the volume of communication (number of calls or total duration) w_{ij}. While the number of communication events between a tie $i \to j$ represents the *static* strength, the transmissibility T_{ij} represents the *dynamical* strength of a tie, i.e. the probability that tie $i \leftrightarrow j$ transmits the information given the actual pattern of communication between i and j. The use of one or the other of these quantities as a proxy for the intensity or strength of a tie may lead to significantly different pictures of the topological structure of the network.
This becomes clear in Fig. 5.8, where we compare the structure of our network around a randomly chosen node when the weight of each tie ij is given by the w_{ij}, i.e. total number of $i \leftrightarrow j$ communications in the observation time period T, with the one obtained by considering the *effective weight* w^*_{ij}. This latter quantity is a function of the dynamics of the diffusion process and can be easily recovered from Eq. (5.7), which leads to $w^*_{ij}(\lambda, T) = -T_0/T \ln(1 - T_{ij}(\lambda, T))$. Therefore, contrary to the static and aggregated w_{ij}, the effective weight w^*_{ij} also accounts for the temporal aspects of tie interaction, since it is obtained from the real transmissibility T_{ij} of the tie which incorporates not only the volume of communication w_{ij}, but also all the temporal inhomogeneities of human interaction (see Eq. (5.5)).

5.6 Discussion

In the last years there has been an increasing interest in understanding how the complex topological structure of the underlying contact network affects the diffusion of information, innovation or opinions. However, most of these studies neglect the temporal dimension of human communication as the fact that humans acts in bursts or cascades of events, most of ties form and decay within the observation time period and there are correlations between communication events. In this chapter we have analyzed the impact that these temporal inhomogeneities of communication have in information spreading phenomena. Accordingly to previous studies, we observed that the scaled distribution of relay time $P(\tau)$, that is the time it takes for one user to pass an information to one of his contacts, is heavy-tailed as a reflection of the bursty human activity (Barabási 2005; Karsai et al. 2011; Vázquez et al. 2006). However, $P(\tau)$ also displays a large number of short relay times than the one expected for a homogeneous Poisson process, which can be interpreted as an indication of group conversations, where calls trigger other calls which are, therefore, correlated to one another.

We then analyzed the role that both the bursty nature and the correlated contact sequences have in spreading processes. To this end we simulated the SIR (Susceptible-Infected-Recovered) model on a large data set of mobile telephone call records on a long time period. This allowed us to simulate a spreading process by taking into account both topological and temporal aspects of human interaction. We found that the bursty behavior and event correlations are the two main dynamical ingredients in the understanding of the information diffusion in social networks which have two opposite effects in information reach: while bursts hinder propagation at large scales, conversations favor local rapid cascades. To explain the observed behavior also from a quantitative point of view, we mapped the SIR model into a static percolation process in which social ties are described by the transmissibility, a quantity that measures the probability that an information is transmitted through a given tie. By means of tie transmissibility, we also proposed an effective way to to describe dynamical networks through a static representation of the social network by means of the dynamical strength of ties, a quantity that encompasses both the topological and temporal patterns of human communication.

Finally, by considering a reference null model in which the burstiness of communication is destroyed while the tie creation/removal is maintained, we have shown that also the tie dynamics studied in details in Chap. 3, plays its role in diffusion processes, by slowing down the spreading dynamics.

Our results indicate that, as well as topological properties, also temporal patterns of human interaction play a crucial role in the description and modeling of real social networks and dynamical phenomena such as information spreading. At the same time, our findings also suggest a simple way to model social networks as dynamical entities instead than static objects, by providing a more realistic picture of the systems they represent. Indeed, we have shown that the network structure significantly change when, instead than the aggregated strength, the dynamical strength of tie is considered. Not only the network structure, but also the role and/or functionality of its members is affected when taking into account temporal properties. We have seen, for example, how static hubs (users with a large number of connections) are not necessarily dynamical hubs, i.e. the most effective nodes in transmission processes as it is, instead, traditionally assumed.

References

Adar E, Zhang L, Adamic L (2004) Implicit structure and the dynamics of blogspace. In: Workshop on the weblogging ecosystem

Anderson RM, May R (1992) Infectious diseases in humans. Oxford University Press, Oxford

Aral S, Muchnik L, Sundararajan A (2009) Distinguishing influence-based contagion from homophily-driven diffusion in dynamic networks. Proc Natl Acad Sci USA 106(51):21544–21549

Aral S, Walker D (2011) Creating social contagion through viral product design: a randomized trial of peer influence in networks. Manage Sci 57:1623–1639

Aral S, Walker D (2012) Identifying influential and susceptible members of social networks. Science 337:337–341

Bailey NTJ (1975) The mathematical theory of infectious diseases and its applications. Hafner Press, New York

Bakshy E, Rosenn I (2012) The role of social networks in information diffusion. arXiv:1201.4145v2

Barabási A-L (2005) The origin of bursts and heavy tails in human dynamics. Nat Sci Rep 435:207–211

Barrat A, Barthélemy M, Vespignani A (2008) Dynamical process on complex networks. Cambridge University Press, Cambridge

Barthélemy M, Barrat A, Pastor-Satorras R, Vespignani A (2004) Velocity and hierarchical spread of epidemic outbreaks in scale-free networks. Phys Rev Lett 92:178701

Boguña M, Pastor-Satorras R (2002) Epidemic spreading in correlated complex networks. Phys Rev E 66:047104

Boguña M, Pastor-Satorras R, Vespignani A (2003) Epidemic spreading in complex networks with degree correlations. In: Pastor-Satorras R, Rubi M, Diaz-Giulera A (eds) Statistical mechanics of complex networks, pp 127–147. Springer, Berlin

Breuer L, Baum D (2005) An introduction to queueing theory. Springer, New York

Cebrián M, Frías-Martínez E, Hohwald H, Lara R, Oliver N (2009) How did you get to know that? A traceable word-of-mouth algorithm. In: Proceeding CSE '09 proceedings of the 2009 international conference on computational science and engineering, vol 04, pp 292–297

Daley D, Kendall D (1964) Epidemics and rumors. Nature 204:1118

Danon L, Arenas A, Díaz-Guilera A (2008) Impact of community structure on information transfer. Phys Rev E 77:036103

Díaz-Guilera A, Sergio L, Arenas A (2009) Propagation of innovation in complex patterns of interaction. In: Pyka A, Scharnhorst A (eds) Innovation networks: new approaches in modeling and analyzing. Springer, Berlin

Eckmann J-P, Moses E, Sergi D (2004) Entropy of dialogues creates coherent structures in e-mail traffic. Proc Natl Acad Sci USA 40:14333–14337

Eguiluz V, Klemm K (2002) Epidemic threshold in structured scale-free networks. Phys Rev Lett 89:108701

Goffman W, Newill V (1964) Generalization of epidemic theory: an application to the transmission of ideas. Nature 204:225

Grabowicz P, Ramasco J, Moro E, Pujol J, Eguiluz V (2012) Social features of online networks: the strength of intermediary ties in online social media. PLoS ONE 7(1):e29358

Guardiola X, Díaz-Guilera A, Pérez CJ, Arenas A, Llas M (2002) Modeling diffusion of innovation in a social network. Phys Rev E 66:026121
Harris TE (2002) The theory of branching processes. Springer-Verlag, Berlin
Hidalgo C, Rodriguez-Sickert C (2008) The dynamics of a mobile phone network. Phys A 387:3017
Holme P, Newman MEJ (2006) Nonequilibrium phase transition in the coevolution of networks and opinions. Phys Rev E 74:056108
Iribarren J, Moro E (2009) Impact of human activity patterns on the dynamics of information diffusion. Phys Rev Lett 103:038702
Iribarren J, Moro E (2011b) Branching dynamics of viral information spreading. Phys Rev E 84:046116
Isella L, Stehlé J, Barrat A, Cattuto C, Pinton J-F, Van den Broeck W (2011) What's in a crowd? Analysis of face-to-face behavioral networks. J Theor Biol 166:166–180
Karsai M, Kivelá M, Pan R, Kaski K, Kertész J, Barabási A-L (2011) Small but slow world: how network topology and burstiness slow down spreading. Phys Rev E 83:025102(R)
Kenan E, Robins JM (2007) Second look at the spread of epidemics on networks. Phys Rev E 76:036113
Kitsak M, Gallos LK, Havlin S, Liljeros F, Muchnik L, Stanley H, Makse HA (2010) Identification of influential spreaders in complex networks. Nat Phys 6:888–893
Kossinets G, Watts DJ (2006) Empirical analysis of an evolving social network. Science 311:5757
Kossinets G, Watts DJ (2009) Origins of homophily in an evolving social network. Am J Sociol 115(2):405–450
Lazer D, Pentland A, Adamic L, Aral S, Barabási A-L, Brewer N, Christakis NA, Contractor N, Fowler J, Gutmann M, Jebara T, King G, Macy M, Roy D, Van Alsyne M (2009) Computational social science. Science 323:721–723
Leskovec, J, Backstrom L, Kleinberg J (2009) Meme-tracking and the dynamics of the news cycle. In: KDD '09 proceedings of the 15th ACM SIGKDD international conference on knowledge discovery and data mining, pp 497–506
Liben-Nowell D, Kleinberg J (2008) Tracing information flow on a global scale using internet chain-letter data. Proc Natl Acad Sci USA 105(12):4633
Lloyd AL, May R (2001) How viruses spread among computers and people. Science 292:1316–1317
Malmgren RD, Stouffer DB, Motter AE, Amaral LAN (2008) A poissonian explanation for heavy tails in e-mail communication. Proc Natl Acad Sci USA 105:18153–18158
Manski C (1993) Identification of endogenous social effects: the reflection problem. Rev Econ Stud 60(3):531–542
McPherson J, Smith-Lovin L, Cook J (2001) Birds of a feather: homophily in social networks. Ann Rev Sociol 27:415–444
Miritello G, Lara R, Moro E (2013) Time allocation in social networks: correlation between social structure and human communication dynamics. In: Holme P, Saramäki J (eds) Temporal networks. Springer
Miritello G, Moro E, Lara R (2011) Dynamical strength of social ties in information spreading. Phys Rev E 83:045102(R)
Moreno Y, Pastor-Satorras R, Vespignani A (2002) Epidemic outbreaks in complex heterogeneous networks. Eur Phys J B 26:521–529
Moreno Y, Gómez J, Pacheco A (2003) Epidemic incidence in correlated complex networks. Phys Rev E 68:046220
Murray J (1993) Mathematical Biology. Springer, Berlin
Newman MEJ (2002b) The spread of epidemic disease on networks. Phys Rev E 66:016128
Newman MEJ (2003b) The structure and function of complex networks. SIAM Rev 45:167–256
Onnela J-P, Saramäki J, Hyvönen J, Szabó Z, Lazer D, Kaski K, Kertész J, Barabási A-L (2007) Structure and tie strengths in mobile communication networks. Proc Natl Acad Sci USA 104:7332
Pan R, Kivelä M, Saramäki J, Kaski K, Kertész J (2011) Using explosive percolation in analysis of real-world networks. Phys Rev E 83:046112

Park Y, Moore C, Bader J (2010) Dynamic networks from hierarchical Bayesian graph clustering. PLoS ONE 5:e8118
Pastor-Satorras R, Vespignani A (2001a) Epidemic dynamics and endemic states in complex networks. Phys Rev E 63:066117
Pastor-Satorras R, Vespignani A (2001b) Epidemic spreading in scale-free networks. Phys Rev Lett 86:3200–3203
Rogers E (1995) Diffusion of innovations. Free Press, New York
Rybski D, Buldyrev SV, Havlin S, Liljeros F, Makse HA (2009) Scaling laws of human interaction activity. Proc Natl Acad Sci USA 106:12640
Saramäki J, Kaski K (2005) Modeling epidemics with dynamic small-world networks. J Theor Biol 234:413–421
Shalizi C, Thomas A (2011) Homophily and contagion are generically confounded in observational social network studies. Sociol Methods Res 27:211–239
Vázquez A, Oliveira J, Dezsö Z, Goh K-I, Kondor I, Barabási A-L (2006) Modeling bursts and heavy tails in human dynamics. Phys Rev E 73:036127
Vázquez A, Rácz B, Lukács A, Barabási A-L (2007) Impact of non-poissonian activity patterns on spreading processes. Phys Rev Lett 98:158702
Watts DJ, Strogatz SH (1998) Collective dynamics of 'small-world' networks. Nature 393:440–442
Wu T, Zhoud C, Xiaob J, Kurthsa J, Schellnhubera H (2010) Evidence for a bimodal distribution in human communication. Proc Natl Acad Sci USA 107:18803
Zhao Q, Oliver N (2010) Communication motifs: a novel approach to characterize mobile communications. In: NetMob2010

Chapter 6
Conclusion, Contributions and Vision for the Future

6.1 Overview and Conclusion

The work done in this thesis aimed to take a first step toward the emerging topic of how to model dynamically real social networks. We addressed this with the belief that a comprehensive description of social systems and a deep understanding of real phenomena is possible only by taking into account both the topological and the temporal properties of human interactions.

In view of this, our objective has been twofold. On one hand, we aimed to characterize the *topological* and *temporal* properties of humans micro-scale behavior and interactions to improve the current description of a social network and its elements. At the same time, we were interested in understanding and modeling the global consequences that these micro-scale interactions have at a larger scale, focusing in particular on the dynamics of spreading phenomena. To this end we have analyzed the human communication patterns from sensors that have become an integral part of our daily lives: mobile phones. Usually, the analysis of social networks is restricted to the analysis of egocentric, smaller social networks and/or shorter periods of time which makes it difficult to asses the universality of the observations and their generalization to other situations. However, the longitudinal and cross-sectional nature of the empirical data sets that we examined (almost 10 billion communication events between 20 million individual over a period of 19 months) allowed us to analyze the global features of human behavior, characterize phenomena that are typically invisible at small scales and, at the same time, guarantee the universality of our findings.

The current availability of large electronic data sets on human social interactions collected from mobile phone as well as from e-mail or online social networks records, offers big opportunities to study human behavior and social phenomena at unprecedented levels of scale and time resolution. The analysis of these electronic traces, describing minute-by-minute interactions of entire populations of individuals, recently led to the discovering and measurement of many interesting temporal features of human behavior which have been neglected so far: human relationships

G. Miritello, *Temporal Patterns of Communication in Social Networks*, Springer Theses, DOI: 10.1007/978-3-319-00110-4_6,

are not continuously active, the way in which people perform tasks as well as the interaction between any two individual is *bursty* since long periods of inactivity are separated by intense bursts of activity, there are causal correlations between interaction events, etc.

The evident disagreement between these empirical observations and model prediction suggests that traditional models of social networks, in which the underlying contact network is separated from the dynamical system, need to be revised. All the results presented in this work make a contribution into this direction. For example, we showed that the nature of a social relationship can be better characterized when also the way in which two individuals communicate over time and not only aggregated quantities such as the volume of communication between them, is considered. This more complete description not only gives a more complete understanding and characterization of the observed contact ties, but also helps for example in the prediction of the network structure in the future. Furthermore, we proved that topological quantities, as the social connectivity, are actually dynamical quantities, affected by the ongoing formation and deletion of social connections which continuously change the properties of the individual contact network. Contrary to the perception of ever-growing social connectivity, we observed that individuals exhibit a limited social capacity which limits the number of contacts they can maintain in a given time period. This affects the way in which humans allocate time and attention across their social circle with important consequences in global processes as influence phenomena and information spreading. We showed that the dynamics of all those phenomena where the interaction timing and causal relations are crucial (e.g. opinion formation, information spreading, etc.) can be understood and modeled only by taking into account temporal features of human dynamics, which decisively affect the observed outcomes.

The general picture that emerges from our work can be summarized as follows. The observed inhomogeneities of human temporal patterns of interaction significantly change the current vision of a social network from a very stable to a highly complex entity. Not only they are essential for a better understanding of the inherent properties of human behavior, but they also play a crucial role in both the description of the underlying social network and the modeling of all those phenomena which are triggered by the way in which people communicate and behave. Since structure and dynamics are tightly coupled, they can not be disentangled in the analysis and modeling of human behavior as, instead, traditional models do.
Next, we give a detailed summary of contributions and our vision for future research.

6.2 Summary of Contributions and Their Implications

Our aim to merge both topological and temporal patterns of human communication to have a complete description of the network, led us to first identify and characterize the main aspects of human dynamics. In particular, according to previous studies, we observed that:

- Nodes and ties are not continuously active and exhibit a non trivial dynamics. In many real social networks individual can join and/or leave the network and the variation in node arrival and leave rates is large. Analogously, over time connections between them form and destroy at a very high pace leading to an ongoing appearance and disappearance of interaction events that continuously changes the structure of the underlying contact network.
- Within the same network, not all the connections have the same importance/role. This is a reflection of real life where people maintain a large number of relationships with a different strength (family, friends, colleagues, acquaintances). This affects the way in which people distribute their communication across their social circle, whose dynamics can change in time.
- Temporal pattern of human communication are bursty: long periods of inactivity are separated by intense bursts of activity. This happens not only in the way a single individual schedules tasks, but also at the level of the interaction between two individuals.
- The interaction between two individuals triggers communication of these individuals with other people. As a consequence, communication events between adjacent ties are time-correlated, a phenomenon that leads to the observation of group conversations and temporal motifs.

In this thesis we investigated the role that these temporal inhomogeneities play in the characterization of the topological properties of the social network and its elements and in the analysis and modeling of dynamical processes. Although the results and the implications usually go beyond a specific area, for the sake of simplicity the contribution has been organized into three main topics: time allocation, link prediction problem and information diffusion in social networks.

1. **Time allocation in social networks** (Chap. 3) *In real life, people do not pay the same attention to all their relationships. Moreover, social interaction always takes time and the limited amount of time available acts as a constraint on the way in which humans allocate it across their social circle. We investigated whether different strategies of communication exist between individuals and how such strategies may depend on the local properties of the individual network, such as the number of contacts or the activity (e.g. time spent on the phone, frequency of communication, etc.). Specifically, by analyzing the network obtained from the mobile phone communication records of 20 million users, we addressed the problem of time allocation from two different perspectives: first we studied the aggregated or static network during an observation period of 11 months and then we analyzed the network as an evolving entity, e.g. instantaneous network, for a longer time period of 19 months.*

    - By analyzing the aggregated network, we found that all individuals distribute their time very unevenly across their social relationships, with a large proportion of calls going to a small number of ties. People with a large social connectivity, on average, also spend more time on the phone in a given time window. However, when compared with those with smaller networks, they do not devote proportionally more time to communicate and thus on average they have weaker ties. This result, that can be related to Dunbar's theory, is a clear manifestation that there is a cognitive limit in the number of people one knows and keeps social contact with. Interestingly, we showed that in contrast to what has been found for face-to-face and Twitter communication, where the threshold value at which the average time dedication starts to decrease is in correspondence of 100–150 connections (Dunbar's number), we found that in mobile networks this limit is smaller (around 40 connections). The observed difference could be attributed to the fact that in phone communication, beside the cognitive and temporal limits, also monetary constrains play their role or to the fact that some contacts are seen face-to-face rather than called on mobile phone.

To analyze how people with different social connectivities distribute the time across their social circle, for each user we calculated the *disparity*, which measures how diverse is the flux or weight that passes through a node (in our case the communication time per tie). The disparity has been widely used to measure the local heterogeneity of different types of egocentric networks, but to the best of our knowledge it has not been examined in large communication networks. Nevertheless, despite its wide use and applications, we demonstrated its ineffectiveness to assess the individuals' time allocation strategies. Specifically we showed that the disparity is significantly affected and biased by the long tailed nature of the distribution of tie weights. As a consequence, a high value of the connectivity translates into an higher probability to have more weak ties belonging to the tail of the distribution, which results in a large disparity, thus in a more heterogeneous social network.

- We took a big step toward the understanding of individuals' strategies of time allocation by analyzing the *instantaneous* communication network, where the number of social relationships each user keeps contact with, is now calculated at each time instant and not as the total number of interactions revealed in a given time window. This approach allowed us to have a more realistic picture of the network of connections of each individual at any time instant since the ongoing appearance and disappearance of ties over time is now taken into account. Our contributions here has been twofold. First, we developed a procedure to disentangle the real formation/removal of a new/old tie from the tie bursty dynamics. By doing this, we overcome one of the biggest obstacles for the observation and the definition of what constitutes a functional tie: the discrimination of broken ties from large periods of inactivity between active ties. We then determined with high precision the number of removed, created and maintained connections of each individual at any time instant, study the different dynamics of communication, their relations with the network topology and their effects at a global scale.

  When the dynamics of tie formation/decay is taken into account, a different picture emerges. We showed that, contrary to the perception of ever-growing social connectivity, individuals exhibit a social capacity, which limits the number of active connections they can maintain during a certain time period.

In our data set we found that this number, whose average is around 13–15 connections, is significantly smaller than the standard social connectivity, suggesting that this latter usually overestimates the human social capacity. In addition to the social capacity, we also defined the social activity, measured as the rate of change of the ego network. By looking at the unbalance between these two quantities for each user, we showed that dynamical strategies of time allocation do actually exist. Although many individuals have a *normal* or *balanced* social strategy between their capacity and activity, other two interesting classes of strategies emerge: the *social keeping* and the *social wandering* strategies. While individuals of the former class tend to keep a very stable social circle and rarely establish or remove connections, people of the latter class open a very large number of new relationships and close existing ones at a high pace. We also found that such different communication strategies, which do not depend on the number of connections, are instead strictly related to both topological properties (e.g. structural diversity) and individual factors (e.g. age). More interestingly, we demonstrated that such strategies are highly assortative, meaning that social wanderers tend to gather and thus the large volatility around a social wanderer also extend to large proportion of the social network around them. In this picture, the network consists of almost static zones of social keepers surrounded by high volatile clusters of social wanderers. This latter finding has profound consequences for methods of communities identification or influence, since the network around social wanderers changes a lot (60 %) in a brief period of time. We showed that communication strategies also affect information spreading since the choice of one or the other strategy has a significant impact in an individual's capacity to access information which is propagating in the network.

Because of the wide set of novel results and their important implications in terms of both local structure and global phenomena, this analysis actually goes beyond the study of how people distribute their time across its social network. For example, our finding that a static picture of social networks leads to a overestimation of the social connectivity might be crucial in all those situations in which the detection of highly connected people (social hubs) is important, such as the identification of the best spreaders for diffusion and adoption or the optimization of immunization strategies.

2. **Link prediction problem** (Chap. 4) *According to traditional approaches, communication ties have been mainly characterized by the total volume of communication between the involved individuals in a given time period, which defines the strength (or importance) of social relations. In this description, any information about the way in which this volume of communication is distributed within the observation period, is therefore completely neglected. By studying the properties of 700 million communication ties over 19 months, we showed that the same volume of communication may actually correspond to ties with very different temporal pattern. To capture these differences, we proposed and measured some quantities to define the nature of a social tie by taking into account not only the*

volume of communication, but also temporal features such as the level of temporal heterogeneity of the communication or the lifetime of the tie. We showed that identifying ties also by their temporal properties not only leads to a better characterization of a social relationship, but also allows to build successful models for the prediction of tie formation and decay in the near future.

- To capture the properties of temporal aspects of tie communication, we proposed and computed few simple measures which are typically not used to characterize the strength of a tie. Among other quantities, we considered the *(i) average inter-event time* to have a measure of the frequency of communication between two individuals, the *(ii) temporal stability* which tells us about the lifetime of the tie, and the *(iii) coefficient of variation* of the tie inter-event time (time between two consecutive events) distribution, which captures the level of burstiness of communication.
- For each tie in our data set, we calculated the quantities descried above and studied their correlations with the basic topological quantities that traditionally define the strength of a social link, such as the level of reciprocity, number of communication events or topological overlap. We showed that real communication patterns significantly differ from the equivalent Poisson process with the same number of call events, in fact many different tie temporal patterns can be obtained for the same intensity of communication. To mimic the Poisson-like case, we analyzed the same tie properties for a reference model in which all real time-stamps have been shuffled across the whole data set, thus destroying any temporal information about real tie communication.
- We showed that the fact that a tie is not observed in a given time period can be either associated to a period of inactivity (burst) as well as its decay. For this reason it is fundamental to distinguish between *observed* ties, i.e. ties that show at least one communication event within the observation period from *open* ties, i.e. ties observed before and after the observation window but which do not interact within it. We showed that these two definitions yield a significantly different picture of the global network persistence and quantified the difference in terms of tie decay rate.
- We investigated the role of temporal patterns of human communication in the link prediction problem by analyzing the prediction power of many topological and temporal tie metrics. After considering different topological and temporal metrics, we proposed a very simple and efficient criterium to predict with roughly the 80 % of accuracy and 19 % of error whether a tie is likely to be newly-formed or to decay in a short future window. Although a deep analysis is out of the scope of this thesis, we showed that our simple methods gives better performance than more complex models often used to predict tie appearance and decay.

The impact that the bursty pattern of ties communication has on the definition of whether a tie is decayed or not, combined with the mechanism of tie formation/decay, demonstrates that a significantly different picture of the network and its future structure emerges depending on whether the human dynamics is incorporated or not in the analysis of social network. Our results show that a much realistic and complete description of social relationships and network structure can be easily obtained with the help of very simple quantities that capture the main temporal features of tie communication.

3. **Information diffusion** (Chap. 5) *We investigated the temporal patterns of human communication and their influence on the spreading of information in social networks by analyzing the mobile phone network of 20 million users over a period of 11 months. First, we analyzed the temporal patterns of human communication and characterized their main properties such as the burstiness and the existence of group correlations. Then, we investigated their role in the dynamical spreading of information by simulating the simple susceptible-infectious-recovered (SIR) model on the top of the real-time sequence of phone records. By means of the tie transmissibility we were able to give a quantitative explanation of the outcomes of the simulations and demonstrate the way in which real patterns of human communication affect the diffusion process. Finally, we provided a simple way to incorporate the causal and temporal patterns of interaction into an effective static representation of the network. This approach represents a novel and efficient procedure to describe and model dynamical social networks.*

    - We analyzed the temporal properties of human communication that might affect information diffusion. For comparison, we also analyzed the results obtained for the Poissonian reference model in which the time stamps of the whole data records have been shuffled, thus destroying any temporal feature and correlation. Accordingly to previous studies, what emerges is that (*i*) tie communication is bursty and (*ii*) there exist groups of conversation among individuals.
    - We investigated the role that these two properties of human interaction play on information spreading. To this end we simulated the SIR model on the real-time sequence of communication events for a wide range of values for the transmission probability λ. Specifically, first we ensured the existence of a percolation transition by studying the size of the largest information cascade. We then analyzed the temporal dynamics of the average cascade size. The results show the existence of two different regimes depending on the value of the transmission probability. When information propagates easily (large λ), the average extent obtained for the real-time sequence is narrower than the one expected when a Poissonian dynamics is considered. However, for small λ, corresponding to a more realistic case, we observed the opposite behavior: information cascades are larger in real data. The differences observed between the real case and the Poissonian expectations reflect the impact of the real-time dynamics of communication on spreading processes. Since human dynamics is typically modeled as a Poisson-like process, our findings not only help in the understanding of real human dynamics and phenomena but also, and most of all, raise the more general question of how to how to model real social networks.

- To give a quantitative explanation of the observed outcomes and take a step toward a dynamical modeling of real communication networks, we mapped the dynamical SIR model to a static edge percolation process where ties are described by the *transmissibility*, e.g. the probability that the information is transmitted from one individual to the other. Our main motivation for this approach was the expectation that, since the transmissibility is strictly related to the transmission process, a quantitative understanding of this latter quantity could have led to a full description and modeling of the observed behavior. Our contribution adheres to the following three steps. First, we were able to approximate the transmissibility for the two regimes of small and large λ and reduce the complex analytical derivation of the transmissibility to the simpler computation of two quantities that capture the properties of temporal patterns of human communication. Specifically, we showed that for small λ the transmissibility is proportional to the number of interaction events between two individuals within the recovering time window (time window after one of the individuals received the information and is still infected), while for large λ it is given by the complement of the probability of finding zero communication events between them within the recovering time window.
  The fact that these two quantities capture, respectively, the existence of group of conversations and the bursty behavior observed for the real-time communication sequences and discussed above, is the key to understand the second step of our contribution: a quantitative explanation of the observed behavior. We showed that, due to the group correlations, the number of communication events within the recovery time window and thus the average transmissibility, is always larger for the real-time case than for the Poissonian one. This is the reason why for small λ we always observed larger cascades for real communication patterns. In contrast, the bursty nature of human interaction makes the probability of finding zero communication events within the recovery time window larger for the real-time sequences and this is why we found smaller cascades in that region. These results show that both the bursty patterns of human communication and the existence of group conversations are the two main dynamical ingredients in the understanding of the spread of information in social networks. These two effects compete in the spread of information: while correlations between events promote the reach of information, the large periods of inactivity typical of the bursty human behavior hinder it.
  Third, by means of the transmissibility, we measured the secondary reproductive number (e.g. the average number of secondary infections produced by a single individual) for both the real-time and the Poissonian case and for the whole range of λ under consideration. This allowed us to accurately predict the tipping point of the process and the change in the size of the cascades given by the simulations.
- We investigated the impact of the tie creation/removal dynamics on information diffusion processes. To this end we considered a new reference model in which the real time instants at which each communication tie was formed/destroyed is preserved but any other temporal pattern such as the

burstiness or events correlation is destroyed. We then compared the results obtained by running the SIR simulations for this latter case with the ones obtained for the real-time and the Poissonian case discussed above. We showed that the effect of tie dynamics is to slow down the information spreading. This result suggests that, as well as the group conversations and the bursty behavior, also tie dynamics plays a fundamental role in diffusion processes and it has to be taken into account when modeling real phenomena driven by human interaction.

- We suggested a simple way to account for the temporal properties of human communication and still have a static representation of the social network. Specifically, we proposed to use the transmissibility as a measure of the tie strength instead of the intensity, e.g. the total number of interaction events (intensity) between two individuals in the observation time window. Since the transmissibility incorporates not only the volume but also all the temporal inhomogeneities of human interaction, it represents the *dynamical* strength of a tie in contrast to the *static* strength given by the intensity. By calculating the transmissibility for each communication tie in our data set (for several values of infection probability and recovery time) we showed that the *static* and the *dynamical* tie may lead to quite different pictures of the topological structure of the network.

These results indicate the necessity to incorporate temporal patterns of human dynamics in the description of real social networks. Specifically, we believe that the novel vision that we proposed to map the dynamics of human interactions onto a static representation of the social network might help in the understanding and modeling of many areas of network research that are based on information spreading, such as the determination of influence (or centrality) and popularity, community finding and targeting in viral marketing.

6.3 Vision for the Future

We believe that the analysis conducted so far constitutes a step forward into the understanding of how to model dynamical social networks. However, there is a lot yet to be explored and understood and we hope that our results will encourage further investigation on this and other related topics. Based on the presented results, one of our short-term research goals is to investigate whether the properties of network structure may change once the strength of a tie is assigned according not only to the volume of communication between the involved individuals (intensity), but also to the way in which such communication is distributed in time. In fact, as we have seen, ties with the same intensity can be characterized by very different temporal properties, thus when the coefficient of variation of the inter-event time distribution or the tie transmissibility are used to assess tie strengths, a different picture of the network structure could emerge.

The use of a temporal (or dynamical) strength might also lead to a deep understanding of more general phenomena such as the spreading of information, social influence and the formation of communities, and offers new opportunities to reevaluate many results obtained in the literature. For example, consistent with the weak ties hypothesis, the majority of strong ties are found within communities while weak ties act as bridges between these clusters (Granovetter 1973). Weak and strong ties also have a different function within the network: while strong ones are essential to maintain the network's integrity, weak ties favor information diffusion, since they have access to different communities thus different part of the network (Onnela et al. 2007). However, as suggested by Fig. 5.8, ties which are "statically weak" can be "dynamically strong"; this can affect the current picture of social networks and ties functionality. It would be therefore interesting to analyze whether a definition of tie (and tie strength) that accounts for temporal aspects of human communication affects the global structure of the network and the role of its elements. Within this framework, another thing that has not been thoroughly investigated in this study, but we would like to analyze, aims to understand the position that individuals with different communication strategies occupy in the global network and the function and/or role they have in global processes as the diffusion of information. On the one hand, we showed that dynamical social strategies are assortative thus social wanderers tend to cluster with other social wanderers and the same applies for social keepers. On the other hand, we observed that social wanderers tend to access information propagating in the network later than social keepers. Some of the questions that arise in this context are: how do dynamical communication strategies influence propagation? Are social keepers better propagators than social wanderers at a global scale? Which individuals should be chosen as potential initiators to optimize the diffusion of a piece of information, rumor or opinion through the society? In fact, although the results suggest that social keepers would react faster, the asymptotic reach of information spreading might be larger (even if slower) when information propagates through social wanderers, due to the smaller cluster observed in their neighborhood and their capability to reach different zones of the network. This analysis would be in line with a recent study based on Facebook communication (Ugander et al. 2012) which shows that the spread of ideas, fads, innovations, depends on the variety of connections (number of distinct social groups) that hold them, more than the total number of friends. Under this perspective, despite the longer time in which social wanderers receive an information, they could have access to a more diverse range of ideas, news and opinions. Analyzing the relationships between community structure and spread of information and opinions by using mobile phone data, would then extend previous research mostly focused on online settings. Moreover, it would give more insight on how to target people in viral marketing campaigns, understanding and modeling churn or word-of-mouth mechanisms in peer-to-peer networks.

Furthermore, the mechanism itself of community formation and evolution can be significantly affected by considering the temporal patterns of human communication. Many of the most widely used algorithms to identify communities and groups in complex networks are based in fact on a static picture of the underlying network (Girvan and Newman 2002; Palla et al. 2005). However, communities are dynamical objects, they rapidly change in time, form, destroy or continue to exist even after all members have been replaced by new members (Palla et al. 2007). To incorporate the role of time dimension into community formation and study community dynamics, one approach has been to consider aggregated time windows of the temporal networks: first communities are detected in different aggregated time slices, and then compared to determine correspondences (Mucha et al. 2010; Palla et al. 2007; Rosvall and Bergstrom 2010). Other approaches have combined both community structure and their temporal evolution by considering the historic evolution patterns of the community to determine its state at a given time instant (Lin et al. 2008). Within this frame, it would be interesting to explore new clustering algorithms based on time-ordered patterns of real sequences of human interactions, in order to take into account the instantaneous interaction between its members and incorporate all the inhomogeneities of human communication patterns, from the burstiness to the causal correlations between interaction events.

Finally, the rich data sets we used for our analysis allowed us to investigate the structure of large social networks and the way in which the temporal dimension affects and reshape its properties and evolution. An important factor of human behavior has however not been investigated in our analysis, that is the role played by mobility patterns of human dynamics. Much research has been done in the last years to explore how social networks are embedded into the underlying geography (Lambiotte et al. 2008; Leskovec and Horvitz 2008; Scellato et al. 2011), model individual movements (Brookmann et al. 2006; González et al. 2008; Simini et al. 2012; Song et al. 2012) and how they affect the spread of diseases Balcan et al. (2009); Eubank et al. (2004) and to understand the relation between geographic distance and social interactions (Crandall et al. 2010; Cranshaw et al. 2010; Eagle et al. 2009). However, a general understanding of how to merge together topological, temporal and spatial patterns of human interactions to obtain a complete picture of real social systems is still lacking. Nevertheless, recent results (Cho et al. 2011) suggest that a combination of these three aspects of social networks would provide important insights on the nature of human relationships and dynamical phenomena.

In this context, we expect that as well as temporal inhomogeneities, also the spatial properties of human interaction, could affect, among others, all the aspects of social networks analyzed in this thesis. Questions like how likely are individuals who share (or visited) the same geographic location to know each other or how physical location of individuals correlates with the social-network friendship have already been investigated (Crandall et al. 2010; Eagle et al. 2009). However, these questions have been addressed according to a static perspective, where temporal inhomogeneities have been neglected. Furthermore, it has been observed that highly connected social groups tend to span shorter distances than connections bridging together (Volkovich et al. 2012), thus spatial constraints are strictly connected to structural network properties. On the other hand we found that the network topology is strictly correlated with dynamical strategies of human communication and in particular that the neighborhood of social keepers is characterized by stable tight clusters. It would be therefore interesting to analyze the way in which the dynamical communication strategies that emerge from our study depend on the geographical position of the individuals involved. Do people establish stable connections with people who share the same location with them and volatile relationships with others? Are social circles of social keepers and social wanderers concentrated into the same physical place? Due to the implications that such strategies have in terms of network volatility or in spreading phenomena, the answers to these questions could help in the very understanding of not only when and how information and opinion spread across society, but also where it comes from and the spatial range it covers. In this context, we would also like to address these questions for online social networks, where the spatial location could play a different role with respect to the one it plays in offline settings.

A more long-term goal is to analyze the social network built from different sources of data, e.g. mobile-phones, online communication and face-to-face interactions. Although this is still a pipe dream due to the lack of multiple data sources of the same population, this type of analysis would generate powerful insights to understand, model and predict human social systems. It is not clear for example whether the same individual use different channels of communication for different purposes or not. On the basis of our analysis on communication strategies, individuals who behave as *keeper* in their offline network may appear to be *wanderers* in the online one, or vice versa. Due to the correlations that we observed between the strategies and individuals factors, this analysis would allow to analyze how people at different stages of their life spend their time in social interactions and what are their habits of communication.

References

Balcan D, Colizza V, Gonçalves B, Hu H, Ramasco J, Vespignani A (2009) Multiscale mobility networks and the spatial spreading of infectious diseases. Proc Natl Acad Sci U S A 106:21484–21489

Brookmann D, Hufnagel L, Geisel T (2006) The scaling laws of human travel. Nature 439:462–465

Cho E, Myers S, Leskovec J (2011) Friendship and mobility: user movement in location-based social networks. In: KDD'11 proceedings of the 17th ACM SIGKDD international conference on knowledge discovery and data mining, pp 1082–1090

Crandall D, Backstrom L, Cosley D, Suri S, Huttenlocher D, Kleinberg J (2010) Inferring social ties from geographic coincidences. Proc Natl Acad Sci U S A 107(52):22436–22441

Cranshaw J, Toch E, Hong JI, Kitturr A, Sadeh N (2010) Bridging the gap between physical location and online social networks. In: 12th ACM international conference on ubiquitous computing, pp 119–128

Eagle N, Pentland A, Lazer D (2009) Inferring friendship network structure by using mobile phone data. Proc Natl Acad Sci U S A 106(36):15274–15278

Eubank S, Guclu H, Anil Kumar VS, Marathe M, Srinivasan A, Toroczkai Z, Wang N (2004) Modeling disease outbreaks in realistic urban social networks. Nature 429:180–184

Girvan M, Newman MEJ (2002) Community structure in social and biological networks. Proc Natl Acad Sci U S A 99:8271–8276

González M, Hidalgo C, Barabási A-L (2008) Understanding individual human mobility patterns. Nature 453(7196):779–782

Granovetter M (1973) The strength of weak ties. Am J Soc 78:1360–1380

Lambiotte R, Blondel V, de Kerchove C, Huens E, Prieur C, Smoreda Z, Van Dooren P (2008) Geographical dispersal of mobile communication networks. Phys A 387:5317–5325

Leskovec J, Horvitz E (2008) Planetary-scale views on a large instant-messaging network. In: WWW'08 word wide web conference on computer networks, machine learning

Lin Y-R, Chi Y, Zhu S, Sundaram H, Tseng L (2008) Facenet: a framework for analyzing communities and their evolution in dynamic networks. In: Proceedings of the 17th international conference on world wibe web, pp 685–694

Mucha PJ, Richardson M, Macon K, Porter MA, Onnela J-P (2010) Community structure in time-dependent, multiscale, and multiplex networks. Science 328:876–878

Onnela J-P, Saramäki J, Hyvönen J, Szabó Z, Lazer D, Kaski K, Kertész J, Barabási A-L (2007) Structure and tie strengths in mobile communication networks. Proc Natl Acad Sci U S A 104:7332

Palla G, Derényi I, Farkas I, Vicsek T (2005) Uncovering the overlapping community structure of complex networks in nature and society. Nature 435:814–818

Palla G, Barabási A-L, Tamás V (2007) Community dynamics in social networks. Noise Stochast Complex Syst Finan 6601:660106

Palla G, Barabási A-L, Vicsek T (2007) Quantifying social group evolution. Nature 446:664–667

Rosvall M, Bergstrom C (2010) Mapping change in large networks. PLoS ONE 5:e8694

Scellato S, Noulas A, Lambiotte R, Mascolo C (2011) Socio-spatial properties of online location-based social networks. In: ICWSM'11, fifth international AAAI conference on weblogs and social media

Simini F, González M, Maritan A, Barabási A-L (2012) A universal model for mobility and migration patterns. Nature 484:96–100

Song C, Qu Z, Blumm N, Barabási A-L (2012) Limits of predictability in human mobility. Science 327:1018–1021

Ugander J, Backstrom L, Marlow C, Kleinberg J (2012) Structural diversity in social contagion. Proc Natl Acad Sci U S A 109(16):5962–5966

Volkovich Y, Scellato S, Laniado D, Mascolo C, Kaltenbrunner A (2012) The length of bridge ties: structural and geographic properties of online social interactions. In: Proceedings of the sixth international AAAI conference on weblogs and social media

Appendix A
Data and Materials

A.1 Mobile Phone Data Set

The majority of results obtained in this thesis have been obtained by using a large data set of mobile phone communication from a European telecommunication operator in a single country, which name and national market share can not be revealed for data privacy policy. The data consists of the Call Detail Records (CDRs) where at least one of the sender or receiver phone number belongs to the operator under consideration. The records contain the details of any single voice call, short message (SMS) or multimedia message (MMS) services that passed through any device belonging to the operator during the time period under consideration. In order to retain customer anonymity, each number is identified by a key number such that it is not possible to recover the names of the users or their phone numbers. No other information is provided to identify customers thus their privacy and anonymity is totally guaranteed.

Our primary mobile phone calls data included phone calls made or received by roughly 2×10^7 phone numbers over a period of 11 month, from February to December 2009. At a later stage in the study, the set of data has been extended with the CDRs until August 2010 thus the final data set consists of the communication events of almost the same number of users and 1.7×10^9 communication ties over a period of 19 months, from February 2009 to August 2010. Specifically, the former data set has been used to obtain the results in Sect. 3.1 of Chap. 3, while the latter for the rest of the analysis performed in Chap. 5.

A.1.1 *Filtering and Sampling*

In all the work presented in this thesis, we focused exclusively on voice call records, filtering out all other services as SMS and MMS. Although SMS and MMS also represent communication events between two individuals, they could give rise to a contact networks different from the one obtained from voice call records. One of the

G. Miritello, *Temporal Patterns of Communication in Social Networks*,
Springer Theses, DOI: 10.1007/978-3-319-00110-4,

Table A.1 Example of CDR (call detail record) data

Caller	Callee	Timestamp	Duration (s)
2	3	9:10, 11-Aug-08	000753
4	5	9:42, 11-Aug-08	000006
3	2	22:36, 13-Aug-08	000141
4	7	11:10, 15-Aug-08	000029

main difference is due to the nature of the communication channel: while a phone call is undirected, in the sense that it allows communication between the caller and the callee regardless to who initiates the phone call, SMS and MMS records represent directed communication where the information is transmitted only in one direction, from the sender to the receiver. Each data record contains the hashed numbers of the caller and the receiver, the date and time at which the call was initiated and the duration of the call. A schematic representation of four entries of our database is shown in Table A.1.

To translate the phone log data into a network representation that captures the characteristics of the underlying communication network, it is usually necessary to preprocess the data by filtering out any spurious event which may not represent a social relationships. To this end we first filtered out all the incoming or outgoing calls that involve other operators, keeping only those events in which both the caller and the receiver numbers belongs to the operator under consideration. This filtering is justified by the fact that we have partial access to the activity of other providers. It is therefore needed to avoid the bias between the operator under consideration and other mobile service providers as we would have a full access to the customers of our operator, but only partial access to the activity of other providers. Moreover, to avoid business-like subscriptions, which usually appear as users with a huge number of connections and calls never returned, we only retain ties which are reciprocated, if there had been at least one reciprocated pair of phone calls between them. This restriction, which also eliminates calls to wrong numbers, telemarketing-type calls, customer service lines, etc., should ensure that we are dealing with a more realistic network of social interactions. All these filtering lead to the removal of about the 50 % of the total links in our database.

A.1.2 Definition of the Network

Due to the bidirectional nature of phone communication and to the fact that there is no reason to assume *a priori* that the user who initiated the call has different properties or role within the network, we neglect the directionality of links and consider a call from i to j equivalent to a call from j to i. Therefore, in our network, a link between two users i and j corresponds to an undirected link with at least one reciprocated pair of phone calls between them during the whole period under consideration. We have

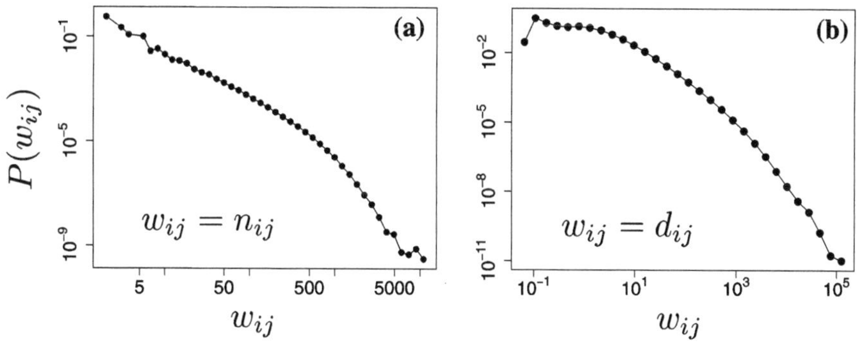

Fig. A.1 Distribution of tie weight w_{ij}. For all ties in our database we show the distribution of tie weight measured as **a** the total number n_{ij} of phone calls and **b** the total duration d_{ij} (in minutes) of phone calls between users i and j

defined the tie weight w_{ij} as (i) the total number n_{ij} or (ii) the total duration d_{ij} of phone communications between i and j during the time period under investigation depending on what we were interested in to investigate. The distribution of both n_{ij} and d_{ij} is shown in Fig. A.1.

Before motivating the choice we have made for each of the cases under consideration, it is important to show that as expected, and in line with other studies on mobile phone networks (Onnela et al. 2007), n_{ij} and d_{ij} are statistically dependent, giving rise for our database to a Person's linear correlation coefficient of 0.689. The relation between these two variable is depicted in Fig. A.2. Because of the proportionality

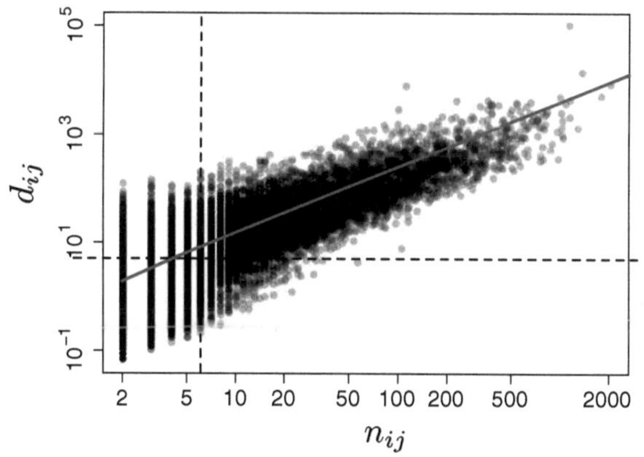

Fig. A.2 Proportionality between the total number and the total duration of phone calls. Total duration d_{ij} (in minutes) as a function of the total number n_{ij} of phone calls between two users i and j during a period of 11 months. The *red line* refers to the linear fit $y = x^{\gamma}$ with $\gamma = 1.2$, while the *black dashed lines* correspond to the median of d_{ij} (*horizontal*) and n_{ij} (*vertical*), which value are, respectively, 6.07 (minutes) and 5 (calls)

between the total number and the total duration of phone calls, the use of either one or the other of these quantities to define the tie weight, leads to an equivalent representation of the network. The main difference is that while n_{ij} is a discrete quantity, d_{ij} can be considered a continuous variable, since in our database it is measured in seconds.
As a consequence, depending on whether the duration of the interaction was important or not for the purpose of the analysis, we have chosen $w_{ij} = n_{ij}$ or $w_{ij} = d_{ij}$ to describe the tie weight. In any case w_{ij} represents a symmetric weight, i.e. $w_{ij} = w_{ji}$, as a consequence of the fact that we are considering undirected links.
For this reason, in Chap. 3 we have defined w_{ij} as the total duration of phone communications between two users, since we were interested in studying the way in which an individual distributes across his social connections the total time he spends on the phone. In this case, in fact, the total time of phone calls reveals much more information than the number of communication events, since it actually captures the temporal (and possibly also the monetary) commitment to the social relationship.
In contrast, in Chap. 5 and in particular in Sect. 5.2.3, we have considered the total number of phone calls between i and j to compute the tie transmissibility associated to the dynamical SIR model. The motivation behind this choice is due to the fact that we simulated the dynamical SIR process on the real-time series of communication events by assuming that the value of the transmission probability λ is fixed and equal for any phone call, regardless to its duration. In this case the duration of phone calls can be therefore neglected, since for the purpose of the analysis, it does not carry any additional information.
Note that in principle, one can also simulate a dynamical process by assuming that the more the duration of a phone call, the higher the chance that a piece of information is transmitted, in which case at each phone call it would be $\lambda(d) \sim d$, being d the duration the phone call. The latter, however, was out of the analysis performed.

As stressed in the whole thesis, although both the n_{ij} and d_{ij} are a good representation of the emotional intensity of a given social relationship (Hill and Dunbar 2003; Wellman and Wortley 1990; Saramäki et al. 2012), one of the goals of the work done has been to show that the aggregated volume of communication does not completely define the underlying relationship between two individuals. In fact, it does not contain any information about the temporal pattern of the tie communication, that is the way in which a given volume of interaction events has been distributed in time.
Specifically, this issue has been discussed in Chap. 4, where we have introduced alternative measures for the definition of the weight of a link. In particular we have defined the *stability* of the link and the *coefficient of variation* of its inter-event time distribution, to capture the lifetime of the link and the burstiness of its communication.
In Chap. 5 we have defined the weight of a link *a posteriori*, accordingly to a dynamical process and in particular to the epidemic SIR model. This led us to the introduction of the tie *transmissibility* as measure of the dynamical weight of a link. Contrary to the static weight obtained when the mere number or duration of phone calls between two users is considered, the transmissibility also capture information on both the properties of the temporal series of communication events between i and j and correlations between these series and communication events with third parties.

A.1.3 Reference Models

In social network analysis, a common way to assess the importance and truthfulness of the obtained results is to compare them against some reference (or null) model where some of the features (i.e. contact network, tie weight, temporal correlation) of the original network have been randomized. For this reason one usually refers to the resulting network as the *randomized network*. In any randomized network, some of the correlations observed in the real network are destroyed in order to understand their contribution to the observed outcomes. Thus, depending on the case under consideration, it can be more convenient to adopt one reference model or another. In particular, in the work done in this thesis three different reference models have been adopted, which are described in the following.

1. *Tie-weight shuffle* The weight of each tie (measured as either n_{ij} or d_{ij}) is replaced by a randomly selected tie weight from the whole network. This shuffle preserves the overall social connectivity of each user (and thus the network topology), while the amount of time each user dedicates to all his connections does not now correspond to the actual value.

   This reference model has been used in Chap. 3 in the analysis of static social strategies of communication.

2. *Time-stamps shuffle* The original series of time-stamps is randomized or randomly reshuffled across the complete CDR, thus destroying any temporal structure and correlations of the original sequence. For this reason, we also refer to this null model also as *Poissonian shuffling*, since it transforms the series of the real time stamps into a Poisson-like process where communication events are homogeneously distributed and there is no causality in the interaction events. Any structural feature of the network, such as the nodes' degree or tie weight is preserved.

   This reference model has been used in Chap. 4 to analyze the temporal features of communication ties and in Chap. 5 to investigate the role of temporal patterns of human communication in information spreading processes.

3. *Intra-tie time-stamps shuffle* The original time sequence of events is reshuffled only within each tie communication. For a given tie the times of the first and the last communication (thus the tie lifetime) are preserved, while the real times at which any other phone call between them has been made do not now correspond to the real ones. The burstiness of communication and the causal correlation between events are destroyed.

   This reference model has been used in Chap. 5 to investigate the role of tie creations and removals in information spreading processes.

A.2 Facebook Dataset

To investigate whether the dynamical social strategies of communication that we find in mobile communication also emerge in online settings, in Chap. 3 we have also analyzed the network obtained from Facebook interaction. In particular, we have studied the public data set of 90.269 users of the New Orleans Network crawled during December 29th, 2008 and January 3rd, 2009 by Vismanath et al. (2009).[1] The data consists of communication events between users through Facebook wall, where each entrance contains two anonymized user identifiers, meaning the second user posted on the first user's wall and the time-stamp at which the messages was posted. Additional information about the data collection, topological properties and temporal evolution of the resulting network can be found in Viswanath et al. (2009).

Contrary to the mobile phone data, the Facebook data is not steady in time, since the database extends over the early days of Facebook growth and thus it shows a growth in the activity over years, which translates in more wall posts and also more users as a function of time (see Fig. A.3). To minimize this effect, in the study of time allocation strategies in Chap. 3 we have chosen only communication events between users that did show any activity in a observation window $\Omega = 212$ days (the time interval between 1000 and 1212 days in the database) and also which were present 20 days before and after Ω. We do not consider the links to be reciprocated in order to have more data accessible for our analysis. With this filter our database contains 125×10^3 communication events of $\sim 10^4$ users and 69×10^3 ties.

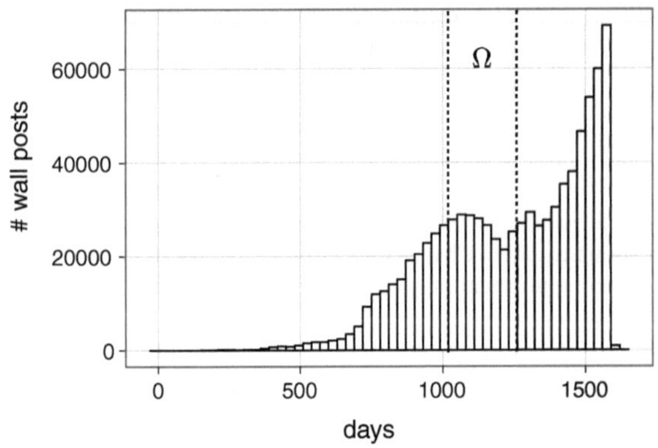

Fig. A.3 Activity in the Facebook database. Total number of communication events through the wall as a function of time for the Facebook database by Vismanath et al. (2009). The vertical dashed lines show the limits of the observation time window Ω used in the analysis of Sect. 3.4.1

[1] Data available at http://socialnetworks.mpi-sws.org.

References

Hill R, Dunbar R (2003) Social network size in humans. Hum Nat Interdisc Biosoc Perspect 14(1):53–72

Onnela J-P, Saramäki J, Hyvönen J, Szabó Z, Lazer D, Kaski K, Kertész J, Barabási A-L (2007) Structure and tie strengths in mobile communication networks. Proc Natl Acad Sci USA 104:7332

Saramäki J, Leicht E, Lopez E, Roberts S, Reed-Tsochas F, Dunbar R (2012) The persistence of social signature in human communication. arXiv:1204.5602

Viswanath B, Mislove A, Cha M, Gummadi K (2009) On the evolution of user interaction in Facebook. In: Proceedings of the 2nd ACM workshop on online, social networks, pp 37–42

Wellman B, Wortley S (1990) Different strokes from different folks: community ties and social support. Am J Sociol 96:558–588

Author Biography

Giovanna Miritello is a Data Scientist at Telefónica Research in Madrid (Spain) in the area of User Modeling and Machine Learning. Her current research lies in the field of social networks with a particular focus on applications ranging from the spread of information and opinions to social influence mechanisms. She received her BSc and MSc in Theoretical Physics from the University of Catania (Italy), where she was a member of the Chaos and Complexity Theoretical University Study (CACTUS) group. From 2008 she has been a member of the GISC (Inter-disciplinary Group of Complex Systems) group at the University Carlos III of Madrid (UC3M), Spain. She was awarded the Ph.D. fellowship *Beca de formation de doctores Telefónica Investigación y Desarollo y Universidad Carlos III de Madrid* to carry out a joint research project between UC3M and Telefónica Research on temporal patterns of communication in massive social networks. She then obtained her Ph.D. in Mathematical Engineering in 2012.

G. Miritello, *Temporal Patterns of Communication in Social Networks*,
Springer Theses, DOI: 10.1007/978-3-319-00110-4,

MIX
Papier aus verantwortungsvollen Quellen
Paper from responsible sources
FSC® C105338

If you have any concerns about our products,
you can contact us on
ProductSafety@springernature.com

In case Publisher is established outside the EU,
the EU authorized representative is:
Springer Nature Customer Service Center GmbH
Europaplatz 3, 69115 Heidelberg, Germany

Printed by Libri Plureos GmbH
in Hamburg, Germany